Mathematical Mind-Benders

Mathematical Mind-Benders

Peter Winkler

A K Peters, Ltd.
Wellesley, Massachusetts

Editorial, Sales, and Customer Service Office

A K Peters, Ltd.
888 Worcester Street, Suite 230
Wellesley, MA 02482
www.akpeters.com

Library of Congress Cataloging-in-Publication Data

Winkler, P. (Peter), 1946–
    Mathematical mind-benders / Peter Winkler.
        p. cm.
    Includes bibliographical references and index.
    ISBN 13: 978-1-56881-336-3 (alk. paper)
    ISBN 10: 1-56881-336-8 (alk. paper)
    1. Mathematical recreations. I. Winkler, P. (Peter), 1946- Mathematical puzzles.
    II. Title.

QA95.W646 2007
510.76--dc22

                                                      2007025196

Cover illustration: Curve and Three Shadows by Afra Zomorodian (see pages 112, 119–120)

The real authors of this book are the people around the world who sent me these wonderful puzzles, including many readers of my previous book and new friends and colleagues in New England. Special thanks are due to the founders of the Albert Bradley Third Century Professorship in the Sciences at Dartmouth.

But whatever part I am entitled to dedicate goes to my parents,

Drs. Bernard and Miriam Winkler,

who must have hoped their first-born would become a useful, contributing member of society, only to watch him grow up to be a mathematician.

Contents

Preface

Mathematics is not a careful march down a well-cleared highway, but a journey into a strange wilderness, where the explorers often get lost.

—W. S. Anglin

This book is for lovers of mathematics, lovers of puzzles, lovers of challenge. Most of all, it is for those who think that the world of mathematics is orderly, logical, and intuitive—and are ready to learn otherwise!

To appreciate the puzzles, and to solve them, it is necessary—but not sufficient—to be comfortable with mathematics. You will need to know what a point and a line are, what a prime number is, and what the probability is of rolling a double-six when you need it. Most importantly, you will need to know what it means to *prove* something.

You will *not* need a professional acquaintance with mathematics. Your computer, calculator, and calculus text can stay in their boxes; but your thinking-cap will have to be on. In some cases, the more courses you have taken in mathematics, the less likely it is that you will find the answer. In some cases, you will read and understand the answer and still not believe it.

The puzzles themselves come from all over the world, and from people of all walks of life. Since publication of my previous puzzle book,[1] many more people have been sending me puzzles, both new and old. I was shocked but delighted to find, a short time before writing these words, that my collection of unpublished puzzles had reached, in both size and quality, what had previously taken me twenty years or so to accumulate.

Readers of my previous book will find some differences. The puzzles themselves lean a bit more toward the surprising; a few, in fact, come from my article "Seven Mathematical Puzzles You Think You Must Not Have Heard Correctly" for the Seventh Gathering for Gardner. I've paid a bit more attention to source than before, with the result that *some* of the information about the origin of the puzzle may actually be correct. Except in the case of puzzles I devised myself, however, all I can promise is "best effort." At the suggestion

---

[1] *Mathematical Puzzles: A Connoisseur's Collection*, A K Peters, Ltd., 2004.

of some readers, I have tried harder when presenting solutions to indicate how they could have been arrived at; but, alas, in many cases I have either failed to do so convincingly or have no idea myself.

The wording of the puzzles, and of their solutions, is my own, and I must take full responsibility for errors and ambiguities. And there will be some, I can assure you.

Puzzles selected for this book are supposed to be elegant and entertaining; to have easy but illuminating solutions that are challenging to find; to embody some mathematical idea, but not to require advanced mathematics to appreciate or solve. Most of all, for this collection especially, they are supposed to confound the intuition and stimulate the brain. Do they all meet all these criteria? Not on your life. But there are some real gems here, any one of which could bring you joy and enlightenment far beyond the meager price of this volume. Check out Curves on Potatoes, p. 2; or Roulette for the Unwary, p. 3; or Love in Kleptopia, p. 9; or Worms and Water, p. 9; or The Faulty Combination Lock, p. 11; or Names in Boxes, p. 12; or Chameleons, p. 22; or Uniformity at the Bakery, p. 23; or Steadfast Blinkers, p. 23; or Red and Blue Dice, p. 23; or Falling Alice, p. 36; or Alice on the Circle, p. 36; or Coins on the Table, p. 51; or Box in a Box, p. 53; or Impressionable Thinkers, p. 67; or Lemming on a Chessboard, p. 67; or Hats and Infinity, p. 91; or Tower of Bricks, p. 93; or Ice Cream Cake, p. 111; or Curve and Three Shadows, p. 112; or Collapsing a Polygon, p. 114; or...

A word on format. The puzzles are organized into chapters for convenience, classified loosely by mathematical area. The solutions are presented at the end of each chapter (except the last); in the hope that readers will think at least a *little* about each puzzle, I did not wish to make it easy to read a puzzle whose solution follows on the page. Information about the background and source of a puzzle is presented with its solution.

These puzzles are hard. You can be justly proud of solving any of them, and in some cases even of just appreciating the solution.

Good luck, and as they say also in the world of mechanical puzzles, happy puzzling!

Peter Winkler

Warm-Ups

Brain (n.) An apparatus with which we think we think.
—Ambrose Bierce (1842–1914), *The Devil's Dictionary*

We begin with some (relatively) easy problems, just to give you a chance to stretch your brain. These require no fancy mathematics, just a bit of logical thinking.

Half Grown

At what age is the average child half the height that he or she will be as an adult?

Bags of Marbles

You have 15 bags. How many marbles do you need so that you can have a different number of marbles in each bag?

Powers of Two

How many people is "two pairs of twins twice"?

Rolling Pencil

A pencil with pentagonal cross-section has the maker's logo imprinted on one of its five faces. If the pencil is rolled on the table, what is the probability that it stops with the logo facing up?

The Portrait

A visitor points to a portrait on the wall and asks who it is. "Brothers and sisters have I none," says the host, "but that man's father is my father's son." Who is pictured?

Strange Sequence

What symbol should come next in the sequence pictured below?

Language Parameter

For Spanish, Russian, or Hebrew, it's 1. For German, 7. For French, 14. What is it for English?

Attention Paraskevidekatriaphobes

Is the 13th of the month more likely to be a Friday than any other day of the week, or does it just *seem* that way?

OK, time to get a *little* more serious.

Fair Play

How can you get a 50-50 decision by flipping a bent coin?

Curves on Potatoes

Prove that, given two potatoes, you can draw a closed curve on the surface of each so that the two curves are identical as curves in three-dimensional space.

You can finish your warm-up with three probability problems; these do require a *modicum* of calculation.

Winning at Wimbledon

As a result of temporary magical powers, you have made it to the singles' finals at Wimbledon and are playing Serena Williams or Roger Federer for all the marbles. However, your powers cannot last the whole match. What score do you want it to be when they disappear to maximize your chances of hanging on to notch an upset win?

Spaghetti Loops

The 100 ends of 50 strands of cooked spaghetti are paired at random and tied together. How many pasta loops should you expect to result from this process, on average?

Roulette for the Unwary

Elwyn is in Las Vegas for a mathematics meeting and finds himself in a casino with some time before the next talk and $105 in his pocket. He saunters over to the roulette table, noting that there are 38 numbers (0, 00, and 1 through 36) on the wheel. If he bets $1 on a single number, he will win with probability 1/38 and collect $36 (in return for his $1 stake, which still goes to the bank). Otherwise, of course, he just loses the dollar.

Elwyn has just enough time to make 105 such one-dollar bets, so he goes ahead with the plan. What, approximately, do you think is the probability that Elwyn will come out ahead? Is it better than, say, 10%?

Sources and Solutions

Half Grown

Parents of young kids will know this one: two years old! (That is, between the second and third birthdays.) Yes, human growth is highly nonlinear. Puzzle suggested by Jeff Steif, of Chalmers University in Sweden.

Bags of Marbles

Fourteen marbles will do the trick. Put an empty bag into a bag containing one marble, then the second bag into a third containing an additional marble, then the third into a fourth containing an additional marble, etc. so that the ith bag contains altogether $i - 1$ marbles (and $i - 1$ bags).

If you didn't think of putting bags in bags—or thought of it but considered it cheating—you would've needed $0 + 1 + \cdots + 14 = 15 \times 7 = 105$ marbles.

Puzzle contributed by Dick Plotz, of Providence RI.

Powers of Two

Eight. It looks like there are four multiples of two in the phrase: "two," "pairs," "twins," and "twice," leading some folks to guess $2^4 = 16$ people. But a twin is only one person. A classic riddle.

Rolling Pencil

My colleague Laurie Snell caught me on this one; did you fall for it too? Sounds like it should be $\frac{1}{5}$, but of course since 5 is odd, the pencil will stop with a face down and an *edge* up. Thus the answer is zero or perhaps $\frac{2}{5}$, depending on your interpretation of "up," but definitely not $\frac{1}{5}$.

The puzzle appears in Chamont Wang's provocative book, *Sense and Nonsense of Statistical Inference* [58].

The Portrait

This one is an *ancient* classic, which appears in Raymond Smullyan's classic *What is the Name of This Book* [55]. Owing to his lack of siblings, "my father's son" can only mean the host himself, hence the portrait is of the host's son.

Strange Sequence

This one was forwarded to me by Keith Cohon, a lawyer with the Environmental Protection Agency. The sequence is intended to represent the beginning of a reversed alphabet, that is, ZYXW, but with the Z turned 90° (either clockwise or counterclockwise) and each successive letter turned an additional 90°. The next symbol would therefore be < or >, representing a sideways V.

Language Parameter

Seven. This curious conundrum was devised by Teena Carroll, a Georgia Tech graduate student, and actually is (slightly) mathematical. The question answered by the intended parameter is what, in the given language, is the first multi-syllabic positive integer?

4

Attention Paraskevidekatriaphobes

Amazingly, this is true, and as far as I can tell was first observed by Bancroft Brown (a Dartmouth math professor, like your author), who published his calculation in the *American Mathematical Monthly* [11]. My present-day colleague Dana Williams is the one who brought this fact to my attention.

It is not hard to verify that in 688 out of 4800 months in the 400-year cycle of our Gregorian calendar, the 13th falls on a Friday. Sunday and Wednesday claim 687 each, Monday and Tuesday 685 each, and Thursday and Saturday only 684 each. To check this you need to remember that years which are multiples of 100 are not leap years unless (like 2000) they are divisible by 400.

The origin of superstition concerning Friday the 13th is usually traced to the date of an order given by King Philip IV of France (Philip the fair), dismantling the Knights Templar.

Incidentally, with some training a person (at least, a person like Princeton's redoubtable John H. Conway) can quickly determine the day of the week of any date in history—even accounting for past calendar glitches. For lazier or more present-time-oriented mortals, a useful fact to remember is that in any year 4/4, 6/6, 8/8, 10/10, 12/12, 9/5, 5/9, 7/11, 11/7, and the last day of February all fall on the same day of the week. (This is even easier to remember if you happen to play craps daily from 9 to 5.) That day of the week is Wednesday for 2007, and advances one each year, two before a leap year.

Fair Play

Flip the bent coin *twice* hoping to get a head and a tail; if the head comes first, call the result HEADS; if the tail comes first, call it TAILS. If the result is two heads or two tails, repeat the experiment.

I was reminded of this puzzle by Tamas Lengyel, of Macalester College; its solution is attributed to the late, great mathematician and pioneer computer scientist John von Neumann, and is in fact sometimes called "von Neumann's trick." It relies on the fact that even if the coin is bent, successive flips are (or at least should be) independent events. Of course, it also relies on it being at least *possible* that the bent coin can land on either side!

If you want to minimize the number of flips to get your decision, the above scheme can be improved upon. For example, if you get HH for the first pair of flips and TT for the second, you can quit and call the result HEADS (obviously, TT followed by HH would then be called TAILS.)

More improvements are possible and in fact an article by Şerban Nacu and Yuval Peres [44] shows how to get the last drop of blood out of the process, minimizing the expected number of flips to get a decision, regardless of the coin's probability of landing heads up.

Incidentally, the problem of extracting unbiased, random bits from various tainted random sources is of major importance in the theory of computing, and the subject of many research papers and significant breakthroughs in recent years.

Curves on Potatoes

Intersect the potatoes! In other words, think of each potato as a ghost and stick one of them into the other. The intersection of their surfaces contains curves on each which fit the bill.

This cute puzzle can be found (among other places) in *The Mathemagician and Pied Puzzler* [5].

Winning at Wimbledon

It sounds obvious that you should ask to be ahead two sets to love (it takes three out of five sets to win the men's), and in the third set, ahead 5-0 in games and 40-love in the sixth game. (Probably you want to be serving, but if your serve is like mine, you might prefer your opponent to be serving the sixth game down 0-40 so that you can pray for a double fault.)

Not so fast! These solutions give you essentially three chances to get lucky and win, but you can get six chances—with three services by you and three by Roger. You still want to be up two sets to none, but let the game score be 6-6 in the third set and 6-0—in your favor, of course—in the tiebreaker.

Amit Chakrabarti of Dartmouth suggested the following improvement, based on the idea that traditionally the complete score of a tennis match includes the game scores of all sets and, if the game score was 6-6, the tiebreaker score as well. Then, you could ask, for example, that the score be 6-0, 6-6 (9999-9997), 6-6 (6-0). The theory here (dubious, granted) is that while you were under your magic spell, your opponent was becoming frustrated and exhausted in the second-set tiebreaker and is now more likely to blow one of the six upcoming match points.

Spaghetti Loops

This is an oldie, passed to me by Dartmouth colleague Dana Williams, and is equivalent to the "Blades of Grass Game" on page 198 of *Martin Gardner's Sixth Book of Mathematical Diversions* [26]. You need to compute the probability of making a loop at each stage, then use "linearity of expectation" to conclude that the expected number of loops is the sum of these probabilities.

When you make your ith knot (of 50), you pick up an end, and of the $101 - 2i$ remaining ends you can choose to tie to this one, only one (the other end of this chain) results in a loop. It follows that the probability that your ith knot makes a loop is $1/(101 - 2i)$, and hence the expected total number of loops is $1/99 + 1/97 + 1/95 \cdots + 1/3 + 1/1 = 2.93777485\ldots$; less than three loops! If the number of initial strands is some large n, the expected number of loops approaches half the nth harmonic number, or about half the natural logarithm of n.

Roulette for the Unwary

I heard this one from Elwyn Berlekamp, during the Seventh Gathering for Gardner. It appeared later in his and Joe Buhler's delightful "Puzzles Column" in *Emissary* [3], Spring/Fall 2006.

The game of roulette favors the bank (quite heavily, compared to the European version which has no 00), and as everyone knows, if you repeat an unfavorable bet often enough, you'll usually end up behind. Here each bet sustains an average loss of $1 - (1/38) \times \$36 = \$1/19$, about a nickel.

However, in this case 105 is not often enough! Elwyn only needs to win three or more times to come out ahead, as he would then collect \$108 for his \$105 investment. His probability of *never* winning is $(37/38)^{105} \sim 0.0608$; of winning exactly once, $105 \times (1/38) \times (37/38)^{104} \sim 0.1725$; twice, $105 \times (104/2) \times (1/38)^2 \times (37/38)^{103} \sim 0.2425$. Thus his probability of coming out ahead is one minus the sum of these, namely 0.5242—more than 1/2!

Of course, this doesn't mean Elwyn has Las Vegas by the throat. You may have noticed that when he *fails* to get three wins he loses at least \$33, a lot more than the \$3 profit he ekes out when he wins exactly thrice (which, given that he does come out ahead, occurs more than 2/3 of the time). *On the average*, Elwyn will end up losing $\$105 \times (1/19) \sim \$5.53$.

For a more extreme example of this phenomenon, suppose Elwyn has \$255 but needs \$256 for a mathematics conference registration fee. His best course of action is to plan on betting \$1, then \$2, then \$4, \$8, \$16, \$32, \$64 and finally \$128 on RED (or BLACK). The first time he wins, he collects double his stake and quits immediately, having exactly the \$256 he needs. He fails only if he loses all 8 bets (and all his money), which happens with probability only $(20/38)^8 < 0.006$.

You can do this yourself. If you can afford to lose \$255 in the worst case, you can enter a casino and get yourself a better than 99% probability of coming out ahead. Then you can quit gambling for the rest of your life. Highly recommended.

Stretching the Imagination

You cannot depend on your eyes when your imagination is out of focus.
—Mark Twain (1835–1910),
A Connecticut Yankee in King Arthur's Court

These puzzles ask you to devise a *plan*. You'll need some serious creativity for some of them!

Love in Kleptopia

Jan and Maria have fallen in love (via the internet) and Jan wishes to mail her a ring. Unfortunately, they live in the country of Kleptopia where anything sent through the mail will be stolen unless it is sent via padlocked box. Jan and Maria each have plenty of padlocks, but none to which the other has a key. How can Jan get the ring safely into Maria's hands?

Worms and Water

Lori is having trouble with worms crawling into her bed. To stop them, she places the legs of the bed into pails of water; since the worms can't swim, they can't reach the bed via the floor. But they instead crawl up the walls and across the ceiling, dropping onto her bed from above. Yuck!

How can Lori stop the worms from getting to her bed?

Comment: Some sort of canopy over the bed perhaps? To prevent the worms from dropping onto the canopy and then crawling around its lip and dropping onto the bed, you could perhaps build a gutter around the canopy. But then the worms could drop onto the edge of the gutter. Hmmm...

Testing Ostrich Eggs

In preparation for an ad campaign, the Flightless Ostrich Farm needs to test its eggs for durability. The world standard for egg-hardness calls for rating an egg according to the the highest floor of the Empire State Building from which the egg can be dropped without breaking.

Flightless's official tester, Oskar, realizes that if he takes only one egg along on his trip to New York, he'll need to drop it from (potentially) *every one* of the building's 101 floors, starting with the first, to determine its rating.

How many drops does he need in the worst case, if he takes *two* eggs?

The Precarious Picture

Suppose that you wish to hang a picture with a string attached at two points on the frame. If you hang it by looping the string over two nails in the ordinary way, as shown in Figure 1, and one of the nails comes out, the picture will still hang (though possibly in a lopsided fashion) on the other nail.

Can you hang it so that the picture falls if *either* nail comes out?

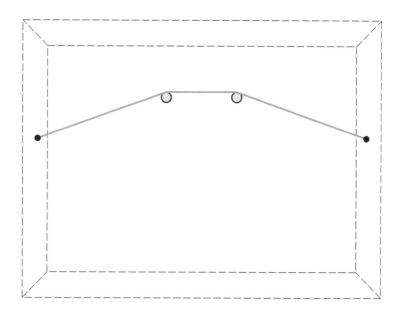

Figure 1. Hanging a picture so that it relies on either nail.

The Faulty Combination Lock

A combination lock with three dials, each numbered 1 through 8, is defective in that you only need to get two of the numbers right to open the lock.

What is the minimum number of (three-number) combinations you need to try in order to be sure of opening the lock?

Comment: There are lots of ways to do this with 64 test combinations, for example you could go through all possible settings of the first two dials, or you could test all combinations whose values sum to a multiple of 8. But each combination you test covers 22 possible cases, and there are only $8^3 = 512$ possible combinations in all, so in theory you might be able to get away with as few as $\lceil 512/22 \rceil = 24$ test combinations. The truth, then, lies somewhere between 24 and 64... but where?

Alternative Dice

Can you design two different dice so that their sums behave just like a pair of ordinary dice? That is, there must be two ways to roll a 3, six ways to roll a 7, one way to roll a 12, and so forth. Each die must have six sides, and each side must be labeled with a positive integer.

Matching Coins

Sonny and Cher play the following game. In each round, a fair coin is tossed. Before the coin is tossed, Sonny and Cher *simultaneously* declare their guess for the result of the coin toss. They win the round if both guessed correctly. The goal is to maximize the fraction of rounds won, when the game is played for many rounds.

So far, the answer is obviously 50%: Sonny and Cher agree on a sequence of guesses (for example, they decide to always declare "heads"), and they can't do any better than that. However, before the game begins, the players are informed that just prior to the first toss, Cher will be given the results of all the coin tosses in advance! She has a chance now to discuss strategy with Sonny, but once she gets the coinflip information, there will be no further opportunity to conspire. Can you help them get their winning percentage up over 70%?

Names in Boxes

The names of 100 prisoners are placed in 100 wooden boxes, one name to a box, and the boxes are lined up on a table in a room. One by one, the prisoners are led into the room; each may look in at most 50 boxes, but must then leave the room exactly as he found it and is permitted no further communication with the others.

The prisoners have a chance to plot their strategy in advance, and they are going to need it, because unless *every single prisoner finds his own name* all will subsequently be executed.

Find a strategy for them that has probability of success exceeding 30%.

Comment: If each prisoner examines a random set of 50 boxes, their probability of survival is an unenviable $\frac{1}{2}^{100} \sim 0.000000000000000000000000000008$. They could do worse—if they all look in the *same* 50 boxes, their chances drop to zero. Thirty percent seems ridiculously out of reach—but yes, you read the problem correctly!

Sources and Solutions

Love in Kleptopia

This puzzle appears in Simon Singh's *The Code Book* [53] and was passed to me by Caroline Calderbank, young daughter of mathematicians Ingrid Daubechies and Rob Calderbank. In the solution she had in mind, Jan sends Maria a box with the ring in it and one of his padlocks on it. Upon receipt Maria affixes her own padlock to box and mails it back with both padlocks on it. When Jan gets it he removes his padlock and sends the box back to Maria; voila! This solution is not just play; the idea is fundamental in Diffie-Hellman key exchange, a historic breakthrough in cryptography.

Depending on one's assumptions, other solutions are possible as well. My favorite was suggested by several persons at the Seventh Gathering for Gardner, including origami artist Robert Lang: it requires that Jan find a padlock whose key has a large hole, or at least a hole which can be sufficiently enlarged by drilling, so that the key can be hooked onto a second padlock's hasp.

Jan uses this second padlock, with the aforementioned key hooked on its hasp, to lock a small empty box that he then sends to Maria. When enough time has passed for it to get there (perhaps he awaits an email acknowledgment from Maria), he sends the ring in another box, locked by the first padlock. When Maria gets the ring box, she picks up the whole first box and uses the key affixed to it to access her ring.

Worms and Water

This curious problem, really as much an engineering puzzle as mathematical, came to me from Balint Virag of MIT.

Lori can indeed do this by hanging a large canopy from the ceiling, extending well over the bed. But the canopy must curve underneath itself at its edges, creating a ring-gutter *underneath* that is filled with water. (See Figure 2 for a cross-section of the canopy.)

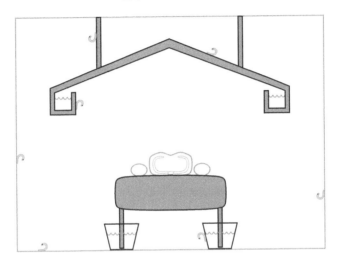

Figure 2. Cross-section of Lori in her worm-proofed bed.

If the worms have no high-up way to get in her bedroom, Lori can more easily accomplish this same task by encircling the room itself with a water-filled gutter.

Testing Ostrich Eggs

A version of this intriguing problem appeared in the delightful book *Which Way Did the Bicycle Go* by Joseph D.E. Konhauser, Dan Velleman, and Stan Wagon [42].

Often it helps to replace a number (here, 102) by a variable, even though you ultimately only care about one value. Let $f(k)$ be the maximum number of floors that you can handle with at most k drops, given that you have two eggs to begin with. So $f(1) = 1$ (to decide if the rating will be 0 or 1). Now suppose Oskar is allowed k drops and takes his first from the nth floor. If the egg breaks, Oskar will have to drop his one remaining egg from floor 1, then 2, and so forth up to $n - 1$, in the worst case; so, the best he can do is to take $n = k$. If the egg survives the drop from floor k, he needs to be able to test all the higher floors with his remaining $k - 1$ drops (using two eggs). It follows that the maximum number of floors Oskar can handle is $f(k - 1) + k$, so we have a recursion: $f(k) = f(k - 1) + k$.

From this we can directly compute that $f(2) = 3$, $f(3) = 6$, $f(5) = 10$, and so forth; in general $f(k)$ is the sum of the numbers from 1 to k. Since there are k such numbers which average $(k + 1)/2$, their total (sometimes called the "kth triangular number") is $k(k + 1)/2$. The first $f(k)$ to reach 102 is $f(14) = 14 \times 15/2 = 105$, so Oskar will need 14 drops in the worst case. Retracing the recursion will tell you how to proceed; for example, the three-floor leeway allows Oskar to drop the first egg from the eleventh, twelfth, thirteenth, or fourteenth floor. Any other choice could cost him an extra drop or leave him eggless without his rating.

To see what happens with three eggs, define $g(k)$ to be the maximum number of floors that can be handled with k drops, beginning with three eggs. Now Oskar can handle $g(k - 1)$ floors above the level of the first drop, if the egg survives that drop; and he can do $f(k - 1)$ floors *below* that level (same f as above) otherwise, because he's down to two eggs now. Hence the new recursion is $g(k) = g(k - 1) + 1 + (k - 1)k/2$, which gives $g(2) = 3$ (no gain yet), but $g(3) = 7$. In general, you can work out that $g(k) = k(k^2 + 5)/6$, and the lowest k for which $g(k) \geq 102$ is 9; thus nine drops are the most Oskar might need from the Empire State Building if three eggs are available.

In general, for large k the number of floors one can handle with m eggs is $k^m/m!$ plus lower order terms; it follows that with m eggs and an n-story super-skyscraper, with n much larger than m, the number of drops needed in the worst case is something like $m! \times n^{1/m}$.

The Precarious Picture

This nice puzzle was contributed by Giulio Genovese, a mathematics graduate student at Dartmouth, who heard it from more than one source in Europe.

One of several ways to hang the picture is illustrated in Figure 3, with slack so you can see better how it works. This solution requires passing the string over the first nail, looping it over the second, sending it back over the first nail, then looping it again over the second nail but this time with a half-twist.

There are also some non-topological solutions: for example, you can pinch a loop of the string between two closely spaced nails, assuming that the nail-

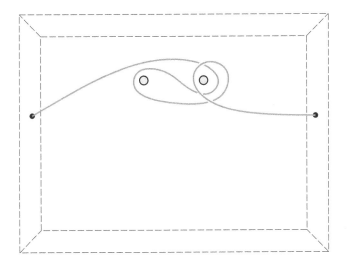

Figure 3. Hanging a picture so that it relies on both nails.

head width is not much larger than the string diameter. But why rely on friction when you can use mathematics?

The Faulty Combination Lock

This very nice combinatorial gem was suggested to me by Amit Chakrabarti of Dartmouth; it had been proposed for the International Mathematics Olympiad in 1988 by East Germany.

Often the easiest way to tackle a problem of this sort is geometrically. The space of all possible combinations is an $8 \times 8 \times 8$ combinatorial cube, and each time you try a combination, you cover all combinations on the three orthogonal lines which intersect at that point.

Once you see the problem this way, you are likely to discover that the best way to cover all the points in the cube is to concentrate all your test points in just two of the eight $4 \times 4 \times 4$ octants. You will then arrive at a solution equivalent to the one described below.

Try all combinations with numbers from $\{1, 2, 3, 4\}$ whose sum is a multiple of 4; there are sixteen of those, since if you pick the numbers on the first two (or any two) of the dials, the number on the third is determined. Now try all the combinations you get by adding (4,4,4), that is, by adding 4 to each of the three numbers; there are 16 more of those, and we claim that together the 32 choices cover all possibilities.

It is easy to see that this works. The correct combination must have either two (or more) values in the set $\{1, 2, 3, 4\}$, or two or more values in the set $\{5, 6, 7, 8\}$. If the former, there is a unique value for the third dial (the one whose number may not be among $\{1, 2, 3, 4\}$) such that the three were among the first 16 tested combinations. The other case is similar.

Here's Amit's way (there are others) to see that we can't cover with 31 or fewer test-combinations. Suppose that S is a cover and $|S| = 31$. Let $S_i = \{(x, y, z) \in S : z = i\}$ be the ith level of S.

Let A be the set $\{1, 2, 3\}$, $B = \{4, 5, 6, 7, 8\}$, and $C = \{2, 3, 4, 5, 6, 7, 8\}$. At least one level of S must contain three or fewer points; we might as well assume S_1 is this level and $|S_1| = 3$. (If $|S_1| < 2$, there's a very easy contradiction.) The points of S_1 must lie in a $3 \times 3 \times 1$ subcube; we may assume that they lie in $A \times A \times \{1\}$.

The 25 points in $B \times B \times \{0\}$ must be covered by points not in S_1. No two of them can be covered by the same point in S, thus, $S - S_1$ has a subset T of size 25 that lies in the subcube $B \times B \times C$. Now consider the set $P = \{(x, y, z) : z \in C, (x, y, 1) \notin S_0, (x, y) \notin B \times B\}$. A quick count shows that $|P| = (64 - 3 - 25) \times 7 = 252$. The points in P are not covered by S_1, and each point in T can cover at most $3 + 3 = 6$ points in P. Therefore, there are at least $252 - 6 \times 25 = 102$ points in P that must be covered by points in $S - S_1 - T$.

However, $|S - S_1 - T| = 31 - 3 - 25 = 3$, and each point in the cube covers exactly 22 points. Since $22 \times 3 = 66 < 102$, we have our contradiction.

Alternative Dice

This problem is famous enough so that its solution has a name: "Sicherman dice." In Martin Gardner's 1978 column [25] or in *Penrose Tiles to Trapdoor Ciphers* [28], you can read about their discovery by Colonel George Sicherman, now of Wayside, New Jersey. Sicherman's unique answer is that the labelings of the two dice are $\{1, 3, 4, 5, 6, 8\}$ and $\{2, 3, 3, 4, 4, 5\}$.

Perhaps you came up with this answer by trial and error, which is a perfectly fine way of solving the problem. However, there is another way in this case, involving a simple example of a powerful mathematical tool called *generating functions*.

The idea is to represent a die by a polynomial in the variable x, in which the coefficient of the term x^k represents the number of appearances of the number k on the die. Thus, for example, an ordinary die would be represented by the polynomial $f(x) = x + x^2 + x^3 + x^4 + x^5 + x^6$.

The key observation is that the result of rolling two (or more) dice is represented by the *product* of their polynomials. For instance, if we roll two ordinary dice, the coefficient of x^{10} in the product (namely, $f(x)^2$) is exactly the number of ways of picking two terms from $f(x)$ whose product is x^{10}. These ways are $x^4 \times x^6$, $x^5 \times x^5$ and $x^6 \times x^4$; and these represent the three ways to roll a sum of 10.

It follows that if $g(x)$ and $h(x)$ are the polynomials representing our ordinary dice, then $g(x) \times h(x) = f(x)^2$. Now, polynomials, like numbers, have unique prime factorizations; the polynomial $f(x)$ factors as $x(x+1)(x^2+x+1)(x^2-x+1)$. To get $g(x)$ and $h(x)$ to multiply out to $f(x)^2$, we need to take each of these 4 factors and assign either one copy of each to $g(x)$ and to $h(x)$, or two copies to one and none to the other. But there's a constraint: we can't have a nonzero number term in $f(x)$ or $g(x)$ (that would represent some sides labeled "0") or any negative term.

The only way to do this (other than $g(x) = h(x) = f(x)$) is to have

$$g(x) = x(x+1)(x^2+x+1) = x + 2x^2 + 2x^3 + x^4$$

and

$$h(x) = x(x+1)(x^2+x+1)(x^2+x-1)^2 = x + x^3 + x^4 + x^5 + x^6 + x^8,$$

or vice-versa.

This may seem much like trial and error after all, but with this technique you can solve problems that are much more complex than this one. To begin with, you can invent alternatives for a pair of eight-sided dice labeled 1 through 8 (there are three new ways to do it), or for rolling *three* ordinary dice (many ways).

Readers who want more depth are referred to an excellent article by Joe Gallian and Dave Rusin [22].

Matching Coins

This problem was brought to my attention by Oded Regev, of the Technion, in Israel.

One way for Sonny and Cher to win more than 2/3 of the time requires dividing up the coinflip sequence into blocks of size three. During each block, Cher "tells" Sonny whether the *next* block is mostly heads or mostly tails; if the former, Sonny guesses "HHH" during the next block; if the latter, "TTT".

How does Cher pass this information? Well, most of the time Sonny will be wrong with (exactly) one of his guesses in the current block, and for that coinflip Cher guesses "H" to tell him the next block is mostly heads, "T" otherwise. For the other two flips in the current round, Cher gives the correct answer (along with Sonny), insuring two out of three wins.

If it happens that Sonny is slated to get all three guesses correct in the current round, Cher uses one of her guesses—say, the third—to communicate as above, even though it might cost them a win on that coinflip.

Thus, after the first block, Sonny and Cher will score two wins out of three whenever the block has two heads and a tail or two tails and a head. When the block is all heads or all tails (which occurs with probability $1/4$), they score two wins out of three half the time and three out of three the remaining half. Altogether this enables them to win $3/4 \times 2/3 + 1/4 \times 5/6 = 17/24 > 70.8\%$ of the time. Note that even if the flips are worst-case (e.g., if the successive heads and tails are called out by an adversary instead of being random) this procedure guarantees a win-probability of at least $1/3$.

Olivier Gossner, Penélope Hernández, and Abraham Neyman [32] have shown that with more sophisticated versions of this scheme, Sonny and Cher can get as close as they want to a fraction x of success, where x is the unique solution to the equation

$$-x \log_2 x - (1 - x) \log_2(1 - x) + (1 - x) \log_2 3 = 1,$$

but they can do no better. Moreover, this applies whether the coinflips are random or adversarial! Since this value of x is about 0.8016, Sonny and Cher can actually arrange to win more than 80% of the time, even when an adversary calls the shots.

Names in Boxes

This puzzle has a short but fascinating history. Devised by Danish computer scientist Peter Bro Miltersen, a version appeared in a prize-winning paper by him and Anna Gal [21]. But Miltersen didn't think there was a solution until one was pointed out to him over lunch by colleague Sven Skyum. Eventually, the puzzle reached me (in a slightly more complex form than presented here) via Dorit Aharonov.

To solve it, the prisoners must first agree on a random labeling of the boxes by their own names. (The point of making it random is that it makes it impossible for the warden to place names in boxes in such a way as to foil the protocol described below.) When admitted to the room, each prisoner inspects his own box (that is, the box with which his own name has been associated). He then looks into the box belonging to the name he just found, and then into the box belonging to the name he found in the second box, etc. until he either finds his own name, or has opened 50 boxes.

That's the strategy; now, why on earth should it work? Well, the process that assigns to a box's owner the name found in his box is a permutation of the 100 names, chosen uniformly at random from the set of all such permutations. Each prisoner is following a cycle of the permutation, beginning with his box and (if he doesn't run over the 50-box limit) ending with his name on a piece of paper. If it happens that the permutation *has no cycles of length greater than 50*, this process will work every time and the prisoners will be spared.

In fact, the probability that a uniformly random permutation of the numbers from 1 to $2n$ contains no cycle of length greater than n is at least 1 minus the natural logarithm of 2—about 30.6853%.

To see this, let $n < k \leq 2n$ and count the permutations having a cycle C of length exactly k. There are $\binom{2n}{k}$ ways to pick the entries in C, $(k-1)!$ ways to order them, and $(2n-k)!$ ways to permute the rest; the product of these numbers is $(2n)!/k$. Since at most one k-cycle can exist in a given permutation, the probability that there is one is exactly $1/k$.

It follows that the probability that there is no long cycle is

$$1 - \frac{1}{n+1} - \frac{1}{n+2} - \cdots - \frac{1}{2n} = 1 - H_{2n} + H_n,$$

where H_m is the sum of the reciprocals of the first m positive integers, approximately $\ln m$. Thus our probability is about $1 - \ln 2n + \ln n = 1 - \ln 2$ and, in fact, is always a bit larger. For $n = 50$ we get that the prisoners survive with probability 31.1827821%.

Eugene Curtin and Max Warshauer [13] have recently shown that this solution cannot be improved upon.

Lambert Bright and Rory Larson, and independently Richard Stanley of MIT, proposed the following variation. Suppose that each prisoner must look in *at least* 50 boxes, and the requirement for survival is that every prisoner does *not* find his own name? Despite having the diametrically opposite objective from before, it seems that the prisoners can do no better than to follow exactly the same strategy. Here, though, they survive only if every cycle has *more* than 50 boxes in it, which can only happen if there is just one big cycle—for which their chances are precisely 1 in 100. Not great, but a lot better than 1 in 2^{100}.

They do just as well even if every prisoner is required to look in 99 boxes—again, they follow the strategy and win just when the random permutation has just one big cycle. In this case it is immediately obvious that no better strategy is available, because the very first prisoner, no matter what he does, has at most a 1% probability of avoiding his own name. The amazing thing is that following the strategy, if that first prisoner succeeds, then automatically every other prisoner will succeed as well!

Numerical Conundrums

The behavior of numbers is a strange and beautiful thing, so it is not surprising
that many wonderful puzzles rely on this behavior and in some cases even
help us to understand it.

Rows and Columns

Prove that if you sort each row of a matrix, then each column, the rows are
still sorted!

Unwanted Expansion

Suppose that you have an algebraic expression involving variables, addition,
multiplication, and parentheses. You repeatedly attempt to expand it using
the distributive law. How do you know that the expression doesn't continue
to expand forever?

Comment: One idea might be to argue that the number of parentheses must go
down, but note that if you apply the distributive law to, say, the outer product
in

$$(x + y)(s(u + v) + t),$$

you get

$$x(s(u + v) + t) + y(s(u + v) + t),$$

which has more parentheses than before.

Chameleons

A colony of chameleons currently contains 20 red, 18 blue, and 16 green individuals. When two chameleons of different colors meet, each of them changes his or her color to the third color. Is it possible that, after a while, all the chameleons have the same color?

Missing Digit

The number 2^{29} has 9 digits, all different; which digit is missing?

A Truly Even Split

Can you partition the integers from 1 to 16 into two sets of equal sizes so that each set has the same sum, the same sum of squares, and the same sum of cubes?

Getting the Numbers Back

For which positive integers n is it the case that, given the $\binom{n}{2}$ pairwise sums of n distinct positive integers, you can recover the integers uniquely?

Evening Out the Gumdrops

In a circle are n children, each holding some gumdrops. The teacher gives an additional gumdrop to each child who has an odd number, then each child passes half of his or her gumdrops to the child on his or her left. These two steps are repeated until they no longer have any effect. Prove that this process will indeed eventually terminate, with all kids having the same (even) number of gumdrops.

The Ninety-Ninth Digit

What is the 99th digit to the right of the decimal point in the decimal expansion of $(1 + \sqrt{2})^{500}$?

Subsets with Constraints

What is the maximum number of numbers you can have between 1 and n, such that no two have a product which is a perfect square? How about if (instead) no number divides another evenly? Or if no two have a factor (other than 1) in common?

Uniformity at the Bakery

A baker's dozen (thirteen) bagels have the property that any twelve of them can be split into two piles of six each, which balance perfectly on the scale. Prove that all the bagels have the same weight.

The next puzzle is a bit more advanced mathematically than most in this collection, but we include it for a special reason.

Anniversary Puzzle

The series $1 - 1 + 1 - 1 + 1 - \cdots$ does not converge, thus the function $f(x) = x - x^2 + x^4 - x^8 + x^{16} - x^{32} + \cdots$ makes no sense when $x = 1$. However, $f(x)$ does converge for all positive real numbers $x < 1$. Does $f(x)$ approach a limit as x approaches 1 from below?

Steadfast Blinkers

Two regular blinkers begin with synchronized blinks at time 0, and afterwards there is an *average* of one blink per minute from the two blinkers together. However, they never blink simultaneously again (equivalently, the ratio of their frequencies is irrational).

Prove that after the first minute (from time 0:00 to time 0:01) there is *exactly one blink* in every interval between time t minutes (t an integer) and time $t + 1$!

Red and Blue Dice

You have two sets (one set red, one blue) of n n-sided dice, each die labeled with the numbers 1 through n. You roll all $2n$ dice simultaneously. Prove that there *must* be a nonempty subset of the red dice and a nonempty subset of the blue dice with the same sum!

Sources and Solutions

Rows and Columns

This is a classical theorem, simple and surprising; I was reminded of it by Dan Romik, now of Hebrew University, Jerusalem. Donald Knuth, in Volume III of *The Art of Computer Programming* [41], traces the result to a footnote in a 1955 book by Hermann Boerner [7]. Bridget Tenner, a student of famed combinatorialist Richard Stanley at MIT, recently wrote a paper called "A Non-Messing-Up Phenomenon for Posets" [57] generalizing this theorem.

To prove Boerner's theorem, let us imagine that the matrix has m rows and n columns, and that after each row is sorted (smallest values to the left, say), the entry on the ith row and jth column is a_{ij}; after the columns are sorted, it is b_{ij}. We need to show that if $j < k$ then $b_{ij} \leq b_{ik}$.

This is one of those things that alternates between being mysterious and obvious, each time you think about it. One way to argue the point is to note that b_{ik} is the ith smallest entry in the old column $\{a_{1k}, a_{2k}, \ldots, a_{mk}\}$. For every entry $a_{i'k}$ that ends up below a_{ik}, the entry $a_{i'j}$ that was on the same row but in the jth column is no larger; thus, counting a_{ij} as well, there are *at least i* entries from the old jth column that are no bigger than b_{ik}. Thus, the ith smallest entry in the old jth column (namely, b_{ij}) is itself no bigger than b_{ik} and we are done.

Is this more convincing than just trying an example? You be the judge.

Unwanted Expansion

This curious puzzle was communicated to me by Dick Lipton, of the College of Computing at Georgia Tech.

One can analyze the expression in terms of depth of trees, but there's an easier way: set all the variables equal to 2! The point of the distributive law is that its application doesn't change the value of the expression. The value of the initial expression limits the size of anything you can get from it by expansion.

Chameleons

Boris Schein, an algebraist at the University of Arkansas, sent me this puzzle; it may be quite old. On one occasion it was given to an eighth grader in Kharkov, on another to a young Harvard grad interviewing at a big finance firm; both solved it!

The key thing is to observe that, after each encounter of two chameleons, the difference between the number of individuals of any two colors remains the same modulo 3. In symbols, letting N_R stand for the number of red chameleons and N_B and N_G similarly, we claim that, e.g., $N_R - N_B$ has the same remainder after division by 3 after any two chameleons meet that it did before. This is easy to verify by checking cases. Thus, these differences remain the same modulo 3 forever, and since in the given colony none of those differences is equal to zero modulo 3, we can never get two of the color populations to be zero.

If, on the other hand, the difference of two color populations (say, $N_R - N_B$) is a positive multiple of 3, we can lower that difference by having a red chameleon meet with a green one (or if there are no green ones, by first having a red one meet a blue one). We repeat until $N_R = N_B$ and then have reds meet blues until only green chameleons remain.

Putting all this together, and noting that if two differences are multiples of 3 then the third must be as well, we see that

- if all three differences are multiples of 3, then any color can take over the colony;

- if just one of the difference is a multiple of 3, then the *remaining* color is the only one that can take over the colony; and finally,

- if none of the differences are multiples of 3, as in the given problem, the colony can never become monochromatic and will remain fluid until other circumstances (e.g., birth, death) intervene.

This puzzle appeared in the International Mathematics Tournament of the Towns (with numbers 13, 15, and 17) of Fall 1984, as Problem 5 in both Junior A-level and Senior O-level. The Tournament of the Towns, from which you will see more puzzles later, was founded in 1980 by Nikolai Konstantinov of Moscow. At the time, the winds of perestroika and glasnost were beginning to blow, and the mathematics competition scene was affected just as every other aspect of Soviet life was. Konstantinov was unhappy with the new measures, resigned from the Central Jury, and formed the Tournament first among small towns in rural Russia. He had gathered around himself a nucleus of outstanding mathematicians, and the success of the Tournament led eventually to Moscow itself becoming one of the "towns." His group also founded the Independent University of Moscow in 1993. Konstantinov's organization has now become the Moscow Center for Continuing Mathematics Education.

The Tournament spread to Poland and Bulgaria, and in 1989, to Australia, through the effort of Peter Taylor of the University of Canberra. Taylor is currently Executive Director of the Australian Mathematics Trust, under whose auspices he has published five books on the Tournament.

In 1990, Andy Liu brought the Tournament to Canada. It has since spread all over the world, with participants from the United States, Western Europe, Asia, and South America. The English version and solutions are prepared by Andrei Storozhev and Andy Liu.

Missing Digit

This amusing riddle was lifted from Berlekamp and Buhler's Puzzles Column in *Emissary* [3], Spring/Fall 2006; they heard it from the number theorist Hendrik Lenstra. You can type "2^29" into Google™ and look at the number for yourself, but is there a way to do it in your head without getting a headache?

Hmm...maybe you remember a technique from grade school called "casting out nines," where you keep adding up digits and always end up with a number's value modulo 9 (that is, it's remainder after dividing by 9). Sometimes called the "Hindu check" or "Arabic check," it uses the fact that $10 \equiv 1$ mod 9, thus $10^n \equiv 1^n \equiv 1$ mod 9 for every n. If we denote by x^* the sum of the digits of the number x, then we get $(xy)^* \equiv x^* y^*$ mod 9 for every x and y.

It follows in particular that $(2^n)^* \equiv 2^n$ mod 9. The powers of 2 mod 9 begin 2, 4, 8, 7, 5, 1, and then repeat; since $29 \equiv 5$ mod 6, 2^n mod $9 \equiv$ the fifth number in this series, which happens to be 5.

Now the sum of *all* the digits is $10 \times 4.5 = 45 \equiv 0$ mod 9, so the missing digit must be 4. Indeed, $2^{29} = 536,870,912$.

A Truly Even Split

This puzzle was suggested by Muthu Muthukrishnan (Rutgers) who heard it from Bob Tarjan (Princeton); both are eminent computer theorists. It turns out that there is indeed such a partition: $\{1, 4, 6, 7, 10, 11, 13, 16\}$ versus $\{2, 3, 5, 8, 9, 12, 14, 15\}$ will do the trick.

To see this, you might think to yourself: Hmmm, 16 is a power of 2; is it possible that this is an example of a more general statement? Can I partition 1 through 8 into two equal-sized sets with the same sum and sum of squares, for example? How about partitioning 1 through 4 into two equal-sized sets with the same sum? The latter is certainly easy: $\{1, 4\}$ versus $\{2, 3\}$. Then, of course, $\{5, 8\}$ versus $\{6, 7\}$ also works for the numbers from 5 to 8, and if you put these together cross-wise, you get $\{1, 4, 6, 7\}$ versus $\{2, 3, 5, 8\}$, which works perforce for sums and now also seems to work for sums of squares.

In general, you can prove by induction that the integers from 1 to 2^k can be partitioned into sets X and Y so that each part has the same sum of jth powers, where j runs from 0 to $k - 1$, and, equivalently, such that for any polynomial p of degree less than k, $p(X)$, which we define as $\sum\{p(x) : x \in X\}$, is equal to $p(Y)$.

To move up to 2^{k+1}, take $X' = X \cup (Y + 2^k)$ (where you get $Y + 2^k$ by adding 2^k to each element of Y) and $Y' = Y \cup (X + 2^k)$. To see that this works, note first that $p(X + 2^k) = p(Y + 2^k)$ since each term is merely another polynomial (q, say) in x_i or y_i. Thus, X' and Y' certainly agree for polynomials of degree less than k, but what if we have a polynomial r of degree k?

But we're OK here too, because the kth power terms on both sides are those of $r(X) + r(Y)$.

Getting the Numbers Back

This puzzle was sent by Nick Reingold of AT&T Labs. The answer if that you can do it if and only if n is not a power of 2! To see this, we again employ the power of polynomials.

Assume first that $X = \{x_1, \ldots, x_n\}$ and $Y = \{y_1, \ldots, y_n\}$ are two different sets with the same pairwise sums. Let $p(z)$ be the polynomial $z^{x_1} + z^{x_2} + \cdots + z^{x_n}$ and $q(z)$ the polynomial $z^{y_1} + z^{y_2} + \cdots + z^{y_n}$. Then, all the cross terms in $p(z)^2$ are the same as those in $q(z)^2$, so $p(z)^2 - q(z)^2 = p(z^2) - q(z^2)$. Dividing through by $p(z) - q(z)$ gives

$$p(z) + q(z) = \frac{p(z^2) - q(z^2)}{p(z) - q(z)}.$$

Now, $p(z) - q(z)$ has 1 as a root (of some positive multiplicity k) since $p(1) = q(1) = n$, so we can write $p(z) - q(z) = (z - 1)^k r(z)$ and similarly $p(z^2) - q(z^2) = (z^2 - 1)^k r(z^2) = (z - 1)^k(z + 1)^k r(z^2)$. Cancelling the factors of $(z - 1)^k$ now gives

$$p(z) + q(z) = \frac{(z + 1)^k r(z^2)}{r(z)},$$

and setting $z = 1$ gives $2n = 2^k$!

To see that you cannot necessarily recover the integers when n is a power of 2, we use the sets X and Y defined recursively for the previous problem. Supposing that X and Y partition $\{1, \ldots, 2^k\}$ and generate the same pairsums, consider $X' = X \cup (Y + 2^k)$ and $Y' = Y \cup (X + 2^k)$. The pair-sums of X' are of the form $x_1 + x_2, y_1 + y_2 + 2^{k+1}$, and $x + y + 2^k$; since those of the form $x_1 + x_2$ are the same as those of the form $y_1 + y_2$ by the induction hypothesis, the pair sums from Y' are exactly the same.

Evening Out the Gumdrops

This problem was submitted by Cliff Smyth (MIT) to "The Puzzle Toad," a delightful puzzle website run by Tom Bohman, Oleg Pikhurko, Alan Frieze, and Danny Sleator at Carnegie Mellon University.[2] It had appeared in the Peking Hgh School Mathematics Competition of 1962 (Grade 12, Paper II, Problem 4).

Let M be the maximum number of gumdrops held by any child at some given time. The first thing to notice is that M cannot increase, unless it is odd, in which case it can increase to the next even number.

To see this, suppose first that M is even; then it is certainly not changed when the teacher adds the "evening" gumdrops, and after passing gumdrops no child can have more than $\frac{1}{2}M + \frac{1}{2}M = M$ gumdrops. Of course, if M is odd it goes up by one to the next even number, and then the previous argument applies when gumdrops are passed.

It follows that after some finite period of time, the teacher will never add another gumdrop. Our task now is to show that after that, the amounts held by the children will "even out."

A nice way to measure the "fairness" of a distribution of a fixed number of items is by considering the sum of the squares of the numbers, which is minimized when the numbers are as equal as possible. Let us consider that parameter in this case, namely $S = G_1^2 + G_2^2 + \cdots + G_n^2$ where G_i is the number of gumdrops held by the ith child. Then, the *change* in S after gumdrops are passed is

$$\left(\frac{1}{2}(G_n + G_1)\right)^2 + \left(\frac{1}{2}(G_1 + G_2)\right)^2 + \cdots + \left(\frac{1}{2}(G_{n-1} + G_n)\right)^2$$

$$- G_1^2 - G_2^2 - \cdots - G_n^2$$

$$= -\frac{1}{2}\left((G_1 - G_N)^2 + (G_2 - G_1)^2 + \cdots (G_N - G_{N-1})^2\right),$$

which is some negative integer (remember, the G_i's are all even now), as long as there's *some* pair of adjacent children with different numbers of gumdrops. Thus, S drops every time gumdrops are passed, until every child has the same number. Since the positive number S cannot keep dropping by whole numbers forever, we are done!

---

[2] http://www.cs.cmu.edu/puzzle.

The Ninety-Ninth Digit

This one is from *Emissary* [3], Fall 1999, and seems difficult. Even if you decide to cheat with your computer, it'll take some patience (or a special program) to extract the information you need.

Instead, note that the expression

$$(1 + \sqrt{2})^{500} + (1 - \sqrt{2})^{500}$$

is an integer because, when expanded, all the odd powers of $\sqrt{2}$ cancel. The right-hand part is very small indeed; in fact, since $|1 - \sqrt{2}|$ is less than $\frac{1}{2}$, and $\frac{1}{2}^5 < .1$, $(1 - \sqrt{2})^{500}$ is far less than 10^{-100} (actually it's about 4×10^{-192}).

Since $(1 + \sqrt{2})^{500}$ is less than an integer by only this minuscule amount, its decimal expansion boasts a huge string of 9's after the decimal point (actually 191 of them, followed by $590591051\ldots$). In particular, the 99th is a 9.

Strange, isn't it? By the same reasoning $(1 + \sqrt{2})^{502}$ is also absurdly close to an integer, yet the ratio of the two giant powers is $(1 + \sqrt{2})^2$, which is a pedestrian $5.82842712\ldots$.

The trick of adding $(1 - \sqrt{2})^{500}$ may seem a bit "out of the blue," but these pairs of conjugate powers, one big and one small, turn up frequently in mathematics. Binet's famous formula

$$F_n = \frac{1}{\sqrt{5}} \left(\frac{1 + \sqrt{5}}{2} \right)^n - \frac{1}{\sqrt{5}} \left(\frac{1 - \sqrt{5}}{2} \right)^n$$

gives you precisely the Fibonacci numbers $1, 1, 2, 3, 5, 8, 13, 21, 34\ldots$, but the second term is small enough so that for any $n \geq 0$, you can just compute

$$\frac{1}{\sqrt{5}} \left(\frac{1 + \sqrt{5}}{2} \right)^n$$

and round it to the nearest integer to get F_n exactly.

Subsets with Constraints

These puzzles can all be tackled the same way. The first was presented by long-time puzzle maven Sol Golomb (University of Southern California) at the Seventh Gathering for Gardner; the second was suggested by Prasad Tetali of Georgia Tech; the third is an obvious variation.

The idea in solving these is to try to cover the numbers from 1 to 30 by bunches of numbers with the property that in each bunch, you can only take one for your subset. Then, if you *can* construct a subset that consists of one choice from each bunch, you have yourself a maximum-sized subset.

In the first case, fix any number k that is square-free (in other words, its prime factorization contains at most one copy of each prime). Now look at the set S_k you get by multiplying k by all possible perfect squares.

If you take two numbers, say kx^2 and ky^2, from S_k, then their product is $k^2x^2y^2 = (kxy)^2$, so they can't both be in our subset. On the other hand, two numbers from *different* S_k's cannot have a square product since one of the k's will have a prime factor not found in the other, and that factor will appear an odd number of times in the product.

Now, *every* number is in just one of these sets S_k—given n, you can recover the k for which $n \in S_k$ by multiplying together one copy of each prime that divides n. Between 1 and 30, the choices for k are 1, 2, 3, 5, 7, 11, 13, 17, 19, 23, 29, 2×3, 2×5, 2×7, 2×9, 2×11, 2×13, 3×5, 3×7, and finally $2 \times 3 \times 5$: 20 in all. You can choose each k itself as the representative from S_k, so a subset of size 20 is achievable and best possible.

To avoid one number dividing another evenly, note that if you fix an odd number j, then in the bunch $B_j = \{j, 2j, 4j, 8j, \dots\}$—that is, j times the powers of 2—you can only take one. If you take for your subset the top half of the numbers from 1 to 30, namely 16 through 30, you have got one from each B_j, and of course no member of this subset divides another evenly since their ratios are all less than 2. So, the 15 numbers you get this way constitute a best possible subset.

Finally, to get a maximum-sized subset all of whose members are relatively prime, you naturally want to look at bunches consisting of all multiples of a fixed prime p. You can take p itself as the representative of its bunch, so you can do no better than to take as your subset all the primes below 30, plus the number 1 itself, for a total of 11 members.

Uniformity at the Bakery

This lovely puzzle is from a Russian competition and appears in *The USSR Problem Book* [51].

Suppose that the weights are not all the same. If the weights are all integers, we can reach a contradiction as follows. Shift the weights down (this does not affect either the weighing property or the conclusion) so that the lightest bagel has weight 0. Now keep dividing all the weights by two until there is a bagel of odd weight. If we leave the odd-weight bagel behind, the sum of the weights of the others must be even, since you can balance them.

But, the same must be true if we leave the weightless bagel behind, and that's impossible.

This argument also works if all the weights are rational numbers, since we can just change units so that the weights are integers. But what if the weights are irrational? It's tempting to replace each weight by a nearby rational number and then proceed as above, but, annoyingly, the previous argument relies on properties of whole numbers and doesn't seem to go through when we try to do it approximately.

It seems that we need to use some serious technology to handle the irrational case. Let us regard the real numbers $\mathbb{R}$ as a vector space over the rationals $\mathbb{Q}$; in other words, think of each real number as a sum of various numbers each multiplied by a rational coefficient. Let V be the (finite-dimensional) subspace generated by the weights of the bagels. Let r be any irrational member of a basis for V, and let q_i be the rational coefficient of r when the weight of the ith bagel is expressed in this basis. Now the same argument used in the rational case shows that all the q_i's must be 0, but this is a contradiction since then r was not in V to begin with.

Incidentally, it is worth noting that the requirement that the weighings have six bagels on a side is necessary. Otherwise, e.g., a baker's dozen consisting of 7 bagels of 50 grams each and 6 bagels of 70 grams each would have the stated property! The place where the proof breaks down is where we shift all the weights down; that relies on there being the same number of items on each side during the weighings.

Anniversary Puzzle

This relatively serious puzzle comes from *Emissary* [3], Fall 2004. It began life, however, in a paper by the great English mathematician Godfrey H. Hardy entitled "On Certain Oscillating Series" [37]. Hardy gets the right answer but comments that no completely elementary proof seems to be known. Luckily for us, the intervening 100 years have given mathematicians a chance to rectify the latter part.

Hardy did not, of course, have access to a computer. If he had tried to solve this puzzle by hand-computing $f(x)$ for various x near 1, he might have reached the wrong conclusion: it *appears* to converge to $\frac{1}{2}$. But appearances can be deceptive, and in fact the limit does not exist.

The following lovely proof comes from the webpage of Noam Elkies, Harvard mathematician and musical composer.[3] Suppose that $f(x)$ does have a

---

[3]http://www.math.harvard.edu/~elkies/Misc/index.html, Problem 8.

limit as x approaches 1 from below.[4] Since $f(x) = x - f(x^2)$, its limit must be $\frac{1}{2}$. But f also satisfies $f(x) = x - x^2 + f(x^4)$, which, since $x^4 < x$, implies that for any c, the sequence $f(c), f(c^4), f(c^{16}), \ldots$ is strictly increasing. It follows that if there is *any* $c < 1$ for which $f(c) \geq \frac{1}{2}$, there can be no limit. In fact, $f(.995)$ (for instance) is $.50088 \ldots$.

What actually happens is that $f(x)$ oscillates more and more rapidly inside an interval of length about $.0055$ centered at $\frac{1}{2}$. Seems a bit capricious, no? The function $g(x) = 1 - x + x^2 - x^3 + x^4 - \cdots$ is also defined for each positive $x < 1$ and has the same problem at $x = 1$ that f did. But this one is equal to $1/(x+1)$, as you can check by adding $xg(x)$ to $g(x)$, thus it docilely approaches $\frac{1}{2}$ as $x \to 1$.

Steadfast Blinkers

This marvelous fact (in another context) was observed by Lord Rayleigh, and probably many others since—and maybe before. One nice modern reference is *Mathematical Time Exposures* by I. J. Schoenberg [52]. Warning: it is a useful fact and may arise again in a later chapter!

To see that this really works, suppose that the first blinker blinks at times pt, $t = 0, 1, \ldots$, and the second at times qt. Then, the frequency of the first is $1/p$ blinks per second, the second $1/q$, and to say that together they average 1 blink per second is to say that $1/p + 1/q = 1$.

To say that the mth blink of the first blinker occurs in the integer time interval $[t, t+1]$ is the same as saying $\lfloor pm \rfloor = t$, where $\lfloor x \rfloor$—the "floor" of x—is the largest integer less than or equal to x. We are trying to prove that every integer can be uniquely represented as either $\lfloor pm \rfloor$ for some integer m, or as $\lfloor qn \rfloor$ for some n, but not both! Can this be true?

Let's first show that we can't get both—in other words, we can't get two blinks in an interval $[t, t+1]$, t an integer. If we could then we have $pm = t + \delta$ and $qn = t + \varepsilon$ for some m and n, where δ and ε are positive quantities less than 1. Let's divide the first equation by p and the second by q, then add to get

$$m + n = \left(\frac{1}{p}t + \frac{1}{q}t \right) t + \frac{1}{p}\delta + \frac{1}{q}\varepsilon \ .$$

But the coefficient of t is just 1, and the quantity on the right is a weighted average of δ and ε—again a positive quantity less than 1. Such a quantity cannot be the difference between two integers, so we have a contradiction.

---

[4]Hardy himself once said: "*Reductio ad absurdum*, which Euclid loved so much, is one of a mathematician's finest weapons. It is a far finer gambit than any chess play: a chess player may offer the sacrifice of a pawn or even a piece, but a mathematician offers the game."

To show that we *do* get a blink in every interval, it's enough to show that we get exactly $t - 1$ blinks after time 0 and before time t; this is because there are no blinks between times 0 and 1, and we have seen that there are at most one in each interval after that. But this is easy, because the first blinker blinks $\lfloor t/p \rfloor$ times in this period, and the second $\lfloor t/q \rfloor$ times. Since $t/p + t/q = t$ and neither t/p nor t/q is an integer, $\lfloor t/p \rfloor + \lfloor t/q \rfloor$ is exactly $t - 1$.

Red and Blue Dice

This wonderful puzzle was brought to me by David Kempe of USC, who needed the result in a computer science paper; it turns out that some similar results can be found in an earlier paper by the very notable mathematicians Persi Diaconis, Ron Graham, and Bernd Sturmfels [14].

One approach to trying to prove the statement is by counting: surely, there are lots of different sums you can get by choosing different subsets of the red dice, and likewise with the blue, so the two sets of sums must overlap? That doesn't seem to fly; for example (choosing $n = 6$ for the moment), you might roll all threes with the reds and all fours with the blues, in which case there are only six possible sums for each color, and it seems somewhat lucky that you can get one sum from each (four red threes, three blue fours) to coincide.

Another temptation is to try induction on n, but that doesn't seem to work either. If you roll no more than one "n" with each set of dice, you can remove a die from each set and reduce to the $n - 1$ case; but if there are multiple n's, you're in trouble.

What to do? Sometimes, paradoxically, you can conquer a puzzle by making it harder. In fact, there is a *much* stronger statement than the one you were asked to prove, which is nonetheless still true. Organize the red dice into a line, any way you want, and do the same with the blue dice. Then, there is a *contiguous nonempty segment* of each line with the same sum.

To put it more mathematically, given any two vectors $\langle a_1, a_2, \ldots, a_n \rangle$ and $\langle b_1, \ldots, b_n \rangle$ in $\{1, \ldots, n\}^n$, there are $j \leq k$ and $s \leq t$ such that $\sum_{i=j}^{k} a_i = \sum_{i=s}^{t} b_i$.

To see this, let α_m be the sum of the first m a_i's and β_m the sum of the first m b_i's. Assume that $\alpha_n \leq \beta_n$ (otherwise we can switch the roles of the a's and b's), and for each m, let m' be the greatest index for which $\beta_{m'} \leq \alpha_m$.

If you wish you can imagine writing the a_i's from left to right, the b_i's beneath them, and drawing a line from each a_m to the rightmost b_ℓ for which the sum of the b_i's up to and including b_ℓ is no greater than the sum of the a_i's up to and including a_m. In Figure 4, two sample strings for $n = 6$ are written out, with the difference $\alpha_m - \beta_{m'}$ written on each connecting line.

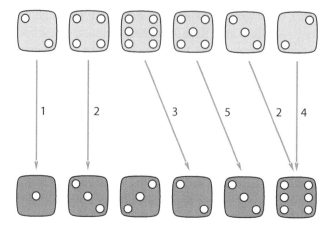

Figure 4. Two "strings" of six ordinary dice, with partial sums matched.

We always have $\alpha_m - \beta_{m'} \geq 0$ and at most $n - 1$ (if $\alpha_m - \beta_{m'}$ were larger than or equal to n, m' should have been a larger index). If any $\alpha_m - \beta_{m'}$ is 0, we are done, as we can then take $j = s = 1$, $k = m$, and $t = m'$ to get two initial segments with the same sum. But, if no $\alpha_m - \beta_{m'}$ is 0 then the n values of $\alpha_m - \beta_{m'}$ all lie in the set $\{1, 2, \ldots, n - 1\}$, so two of them must have the same value. Suppose those are $\alpha_p - \beta_{p'}$ and $\alpha_q - \beta_{q'}$. But then $\sum_{i=p+1}^{q} a_i = \sum_{i=p'+1}^{q'} b_i$, and we are again done.

That was a tricky one, no doubt about it.

In the figure, there is just one pair of coincident differences (both equal to 2), namely with $p = 2$ and $q = 5$. There $p' = 2$ and $q' = 6$, and for our sum-matched substrings we get $a_3 + a_4 + a_5 = 6 + 5 + 3 = 3 + 2 + 3 + 6 = b_3 + b_4 + b_5 + b_6$.

The Adventures of Ant Alice

> Go to the ant, thou sluggard, consider her ways,
> and be wise.
> —Proverbs 6:6–8

Ants, even in a one-dimensional environment, are a source of fascination for amateur puzzlists and mathematicians. We've presented here ten puzzles (devised by this writer, except where noted) involving our "favorite ant" Alice. (See Figure 5.) Each puzzle is intended to illustrate some mathematical idea.

We begin with the basic "ant puzzle."

Figure 5. Ant Alice, in person.

Falling Alice

Twenty-five ants are placed randomly on a meter-long rod; the thirteenth ant from the west end of the rod is our friend, Ant Alice. Each ant is facing east or west with equal probability. They proceed to march forward (that is, in whatever direction they are facing) at 1 cm/sec; whenever two ants collide, they reverse directions. How long does it take before we can be certain that Alice is off the rod?

Alice on the Circle

Now Alice is one of 24 ants randomly placed on a circular track of length 1 meter. Each ant faces randomly clockwise or counterclockwise and marches at 1 cm/sec; as usual, when two ants collide they both reverse directions. What is the probability that after 100 seconds, Alice finds herself exactly where she began?

Guessing the End

The ants are back on the rod. When Alice does fall off, what is the probability that she falls off the end she was originally facing?

Last One Off

What is the probability that Alice is the last ant to fall off the rod?

Counting Collisions

During the process, what is the expected (i.e., average) number of collisions that take place on the rod?

Damage to Alice

What is the expected number of collisions that Alice herself has?

Alice's Insurance Rate

What is the probability that Alice has more collisions than any other ant?

Damage to the Rest of the Ants

Suppose that Alice has a cold, which is transmitted from ant to ant instantly upon collision. How many ants will be infected, on average, before the rod is cleared?

Alice at the Midpoint

Let us do a new experiment. Alice is carefully placed at the exact center of the meter rod, with 12 ants placed randomly to her west and 12 more to her east. As before each ant faces randomly east or west, and they all march in whatever direction they are facing at 1 cm/sec, reversing directions whenever any two meet head-on. This time, however, ants do not fall off the rod; they turn around when they reach the end. One hundred seconds later the ants are frozen in place; what is the maximum distance Alice can be from her initial position?

Alice's New Whereabouts

There are only 24 ants on the rod now, with 12 on the west half facing east and the rest on the east half facing west. Alice is the fifth ant from the west end. They proceed to march as usual, reversing when they collide, and falling off the ends. What do you need to know about the initial configuration in order to predict where Alice will be after 63 seconds?

Sources and Solutions

Falling Alice

As far as I know the first publication of this marvelous puzzle was in Francis Su's "Math Fun Facts" web-column at Harvey Mudd College; Francis recalls hearing it in Europe from someone he can't trace named Felix Vardy. The puzzle then showed up in the Spring/Fall 2003 issue of *Emissary* [3].

Dan Amir, a former Rector of Tel Aviv University, spotted the puzzle in *Emissary* and posed it to Tel Aviv mathematician Noga Alon, who brought it to the Institute for Advanced Study; I first heard it from Avi Wigderson of the IAS, in late 2003.

The key to this (and succeeding) puzzles is to notice that if ants were interchangeable, it would make no difference to the process if they passed one another instead of bouncing. Then it's clear that each ant is simply walking straight ahead and must fall off within 100 seconds. Since all the ants are off in 100 seconds, Alice in particular has fallen off as well.

A nice way to think about the puzzle, which avoids making the ants anonymous, is to imagine that each carries a flag. When two ants meet and bounce, they exchange flags. Thus at all times each ant is carrying *some* flag, and the flags march straight past one another. When all the flags are off the rod, all the ants are off as well.

If you start an ant facing east at the west end of the rod, you can arrange it so that Alice ends up carrying his flag off the east end of the rod 100 seconds later. So, waiting 100 seconds is necessary, as well as sufficient, to be sure the rod is cleared.

Alice on the Circle

As usual, we suppose that each ant carries a flag and that flags are exchanged when two ants meet. Then, each flag travels exactly once around the track in the given time period, ending where it began. The ants themselves must remain in the same circular order in which they started, so they have experienced (at most) some rotation; every one must move the same number of positions, say k positions clockwise. In particular, Alice returns to her initial position only if all the ants do.

Note, however, that if m ants are initially facing clockwise then at any time there are always m ants moving clockwise and $24 - m$ moving counterclockwise. This is because at each collision a clockwise ant becomes counterclockwise and vice versa; or, you can think of it as conservation of angular momentum! In any case, the average ant moves $2m - 24$ cm clockwise during the experiment. Thus, we are back to the initial position if and only if $2m - 24$ is a multiple of 24; i.e., if $m = 0$, 24, or 12.

The first two possibilities (where all ants initially faced the same way) have negligible probability, but the last contributes a healthy 16.1180377%.

To be precise, there are $2^{24} = 16,777,216$ ways to choose directions for the ants, of which $\binom{24}{0} + \binom{24}{12} + \binom{24}{24} = 1 + 1 + 2,704,156$ bring Alice back to where she started. This gives probability $2,704,158/16,777,216 \sim 0.161180377$.

Guessing the End

The number of ants falling off the east end of the rod is the same as the number of ants facing east at the start, since the number of ants facing east never changes. (Alternatively, you can think of flags falling off the rod instead). In any case, if k ants fall off the east end, it is exactly the k easternmost ants who do so, since the ants stay in order.

We may assume, by symmetry, that Alice is facing east at the start, and we know that she goes off the east end just when the number of east-facing ants is at least 13. This means that 12 or more of the *other* 24 ants are facing east. Of course, the probability that 13 or more of 24 ants face east is the same as the probability that 11 or fewer face east, so the probability of the event we are interested in is one-half plus one-half times the probability of exactly 12 of 24 ants facing east. The latter is $\binom{24}{12}/2^{24}$, which works out to 0.161180258; thus, the answer is $0.580590129\ldots$, a bit over 58%.

Last One Off

We may assume (by symmetry, again) that Alice departs by the east end of the rod, which means that the 12 ants to her east do the same. If she is last off, it must be that the 12 ants west of Alice drop off the west end; it follows that initially exactly 12 flags, thus 12 ants, faced west. This happens with probability $\binom{25}{12}/2^{24}$, about 31%.

However, Alice is not necessarily the last ant off in these cases; about half the time, her western neighbor has the honor. Thus, the desired probability is about 15.5%.

But should you be satisfied with an approximation? Not when the exact answer is available. The time each flag is fated to spend on the rod is independent and uniformly random between 0 and 100 seconds. Thus, the probability that the longest-lived flag is one of the 13 east-facing flags is 13/25. It follows that the correct value is $13/25 \times \binom{25}{12}/2^{24}$, which is the same as the now-familiar number $\binom{24}{12}/2^{24}$, about 16.1180258%.

Counting Collisions

Each flag crosses all others ahead of it that are headed toward it; for the average flag, which starts at the midpoint of the rod, this is 6 of the 12 ahead of it. So, the average flag hits six others; thus, there are $25 \times 6 = 150$ "hits" on average. But this counts each collision twice, thus the answer is 75.

An alternate, and slightly more rigorous, way to compute this: What is the probability that two flags cross? No matter where they are, this happens if and only if they face one another, thus with probability 1/4. By linearity of expectation, then, the expected number of flag crossings is $\binom{25}{2} \times 1/4 = 25 \times 24/8 = 75$.

The maximum number of collisions is achieved if all ants are pointed toward Alice (the center ant), in which case all 13 flags facing Alice's way hit all 12 flags facing the other way, for a total of $12 \times 13 = 156$ hits.

The least possible number of collisions is of course zero, but this occurs with probability only $26/2^{25} \sim 0.000000774860382$.

Damage to Alice

It's easy to compute the number of collisions Alice's *flag* has; assuming (say) that Alice initially faces east, there will be an average of 6 (out of 12) flags ahead of Alice facing west, hence her flag expects to pass six other flags.

But Alice is not always carrying her original flag, and in fact we expect Alice to have many more than six collisions on average. Why? Because the *average* ant has six collisions ($75 \times 2/25$), and Alice, being the middle ant, should have more than average.

Now, any given ant collides only with its two neighbors, and alternates between them (since its direction alternates between collisions). An ant's *last* collision will be with its western neighbor if it ends up falling off the east end, and with its eastern neighbor if it goes off the west end.

Suppose that k ants face west initially. Since their flags march off the west end of the rod, the k westernmost ants end up dropping off the west end. Each ant who faced west initially will have an equal number of collisions on each side; those who faced east will have one extra collision on the east side. Thus, the number of collisions between ant j (counting from the west) and ant $j+1$ is equal to the number of east-facing ants among ants 1 through j—as long as $j < k$.

By symmetry, we may assume that k is between 13 and 25 (i.e., Alice herself drops off the west end). Then the number of collisions between Alice and her western neighbor is exactly the number of east-facing ants west of Alice; call this number x. The total number of collisions experienced by Alice would then be $2x$ or $2x + 1$, depending on whether she herself faced west or east at the start.

A priori, the expected value $E[x]$ of x is 6, since there are 12 ants west of Alice and each could face either way. However, we just assumed (darn!) that more than half the ants faced west. Note that since the expected number of east-facing ants *east* of Alice is the same as $E[x]$, the number $2E[x+1]$ that we seek is exactly the total expected number of east-facing ants given that they are in the minority.

Suppose that the ants were assigned directions in alphabetical order, with Ant Zelda last. There are $2^{25}/2 = 2^{24}$ ways to do the assignment so as to get a west-facing majority; of those, $\binom{24}{12}$ result in 12 east-facers among the first 24 choices. In those, Zelda is forced to face west; in the rest she is equally likely to face west or east. It follows that the probability that she faces east is $1/2 - (1/2) \times \binom{24}{12}/2^{24} \sim 0.419409871$.

Since Zelda's probability of facing east is no different from any other ant's, we can multiply this by 25 to get the expected number of east-facing ants, about 10.4852468. This, then, is the average number of collisions experienced by Alice.

Alice's Insurance Rate

Suppose that the westernmost k ants fall off the west end, the rest the east end. We have seen in the previous puzzle solution that, if c_i is the number of collisions between ant i (counting from the west end) and ant $i + 1$, then c_i stays the same or increases by 1 up to $i = k$; after that c_i stays the same or decreases by 1. In particular, $c_i = c_{i-1}$ exactly when ant i faces (initially) the end he or she is fated to drop from.

The number of collisions experienced by ant i is $c_{i-1} + c_i$, so in order for Alice to win the collision game, we need $c_{11} + c_{12} < c_{12} + c_{13} > c_{13} + c_{14}$ which means $c_{11} < c_{13}$ and $c_{12} > c_{14}$. This can only happen if $c_{11} < c_{12} = c_{13} > c_{14}$, which requires that $k = 12$ or 13; that Alice faces the end she drops from; and that her two neighbors face *away* from the ends they drop from. This sounds like probability

$$\left(\binom{25}{12} + \binom{25}{12} \right)/2^{25} \cdot \left(\frac{1}{2} \right)^3 \sim 3.87452543\%,$$

but the events are not quite independent.

Suppose that Alice faces east, her eastern neighbor is Ed, and her western neighbor is Will. Then Ed, like Alice, will be dropping off the east end and therefore must have faced west initially (probability 1/2). Will will be one of the 12 ants dropping off the west end and hence began facing east (probability 1/2). The remaining 22 ants must have begun half facing west, half east (probability $\binom{22}{11}/2^{22}$), so the accurate answer is

$$\left(\frac{1}{2} \right)^2 \cdot \binom{22}{11}/2^{25} \sim 4.20470238\%.$$

Damage to the Rest of the Ants

Like many of the Alice Ant puzzles, this one is purely combinatorial; for example, somewhat counter-intuitively, it has nothing to do with the length of the rod. You might think a shorter rod could enable some ants to get off the rod before they have a chance to become infected, but once an ant is headed for the end with no oncoming ants ahead, its collision days are over.

Probably the easiest way to make the required calculation is to think of flags being infected instead of ants. We may assume that Alice faces east; then all west-facing flags ahead of her will cross hers and become sick, while east-facing flags ahead of her will escape uninfected. In the meantime, the west-facing flags, after crossing Alice's flag, will infect all east-facing flags *behind* Alice while the west-facing flags behind Alice get away.

Since there are an average of 6 west-facing flags ahead of Alice and 6 east-facing flags behind her, this seems to give an average of 13 infected flags (counting Alice's) and thus 13 infected ants.

However, there's a slight glitch: if there are *no* west-facing ants ahead of Alice then there is no flag to cross Alice's and infect the east-facing flags behind her. This happens with probability $1/2^{12}$ and reduces the expected number of infectees from 7 (Alice plus an average of 6 east-facing flags behind her) to 1 (Alice alone), in that case. So the correct answer is not 13 but $13 - 6/2^{12} \sim 12.9985352$ sick ants on average.

Alice at the Midpoint

This puzzle was contributed by John Guilford, of Agilent Inc., to Stan Wagon, who made it the Macalester College "Problem of the Week"[5] at some point in fall of 2003. I heard it from Elwyn Berlekamp, at the Joint Mathematics Meetings in Phoenix, January 2004. It was there that the central character in this chapter received her name; I believe Elwyn actually has an Aunt Alice. I was influenced personally by the presence at the conference of Alice Peters of A K Peters, Ltd., publisher of this book.

Suppose as usual that each ant carries a flag and that flags are exchanged when two ants meet. Then, each flag travels exactly one meter, bouncing once off the end of the rod and ending at a position symmetrically opposite its initial position. In particular, Alice's flag ends up back in the center. But will Alice be carrying it?

Indeed she will, because the ants remain in their original order. The 12 flags originally on the west side of the rod are now on the east side and vice versa, so Alice's flag is once again the 13th flag and Alice herself is still the 13th ant.

So Alice ends exactly where she began; in other words, the maximum distance she can be from her starting spot is zero.

---

[5]http://mathforum.org/wagon/fall03/p996.html.

Alice's New Whereabouts

This is a variation of a puzzle that was devised by Noga Alon and Oded Margalit, of Tel Aviv University, and communicated to me by Noga.

Let $x_1, \ldots, x_{12}$ be the initial positions of the 12 west-facing ants, numbered from west to east; the positions are measured in centimeters from the west end of the rod. Alice's spot is then x_5. Let k be such that the flags beginning at $x_k, \ldots, x_{12}$ remain on the rod, ending, therefore, at $x_k - 63, \ldots, x_{12} - 63$.

The ants, of course, remain in order. Since k flags drop off going west, Alice is gone from the rod if $k \geq 5$. Otherwise she is the $(5 - k)$th remaining ant, counting from the west end, which puts her in position $x_{k+(5-k)} - 63 = x_5 - 63$.

Thus, all you need to know is the position of the 5th ant east of Alice, i.e., the 17th ant from the west end. Alice will end up 63 cm west of that spot; if that spot was already less than 63 cm from the west end, she falls off the rod.

Figure 6. Ant Alice waves "Goodbye."

A Wordy Digression: The Game of HIPE

How every fool can play upon the word!

—Lorenzo, in Shakespeare's *The Merchant of Venice*

This is a free chapter—my publisher assures me that the book would be the same price without it. Think of it as an intermission, a non-mathematical break. Many mathematicians *do* love word games, though, and (in my experience) this one in particular.

You may have heard the following puzzle: what English word contains four consecutive letters that are consecutive letters of the alphabet? Answer: undeRSTUdy. Inspired by this and other word puzzles, I and three other high-school juniors[6] at a 1963 National Science Foundation summer program began to fire letter combinations at one another, asking for a word containing that combination.

The combination had to appear in that order and with no other letters in between. Example: **WKW** = aWKWard, **NSW** = aNSWer. Some were only two letters, e.g., **HQ** = eartHQuake; **ZV** = rendeZVous (or another borrowed word, mitZVah). Try **BV** yourself (hint: this one is obvious).

Double letters (**WW, VV, HH**) can be fun too; but the most deadly combinations we found were three or four letters, as in **GNT, PTC, THAC, HEMU**—answers at the end of the chapter. We named the game after one of our favorite combinations, **HIPE** = arcHIPElago (despite later discovering that "worshiper" can be spelled with one "p"). Of course, HIPE has no doubt been invented and reinvented thousands of times over history, and you are under no obligation to use our name—but it's helpful to call it *something*.

In devising a HIPE, a natural objective is for it to be artfully concealed in the solution; and it's additionally satisfying for the solution to be a common

---

[6]Richard Thurston, Robert Webber, and Robert Winternitz—yes, roommates were assigned alphabetically.

word that's hard to think of. For example, **ONIG** has a solution that is among the most common words in the language: can you find it? And what fun it is to stump a friend with a HIPE like **LYLY**, then tell them they have to add only one more letter.

Ideally, you'd like the solution to be unique (among, say, Scrabble™-eligible, non-capitalized words), but insisting on this leads to unnecessary quibbling. Still, a really good HIPE should have one standout answer. Back in 1963 we agreed on the following HIPE etiquette: when your victim produces a valid solution to your HIPE, his or her obligation is fully discharged. In particular, you are not allowed to ask for the solution you had in mind.

During the summers of 1967 and 1968 I gave a HIPE a day to all the (supposedly) gifted kids at Science and Arts Camps, in Port Ewen, NY. Here are some of their favorites: **SPB, RAOR, XS, DQ, HCR, UDU, YEB, YNG, TIK, XOP,** and **BEK**—answers at end of chapter.

Rarely do good HIPEs of more than four letters arise, since they tend to give away too much information. Two cute ones, notable for their pleasing repetitiveness, are **ACHACH** and **TANTAN** (solutions at end).

Who is good at solving HIPEs? Over the years, I have made an informal study of this question. Of course, as with mathematical puzzles, you never know until the game is played. People you think would be terrific fall flat on their faces, while others shock you by spouting answers you haven't even thought of.

There are some interesting tendencies, though. Once I found an excuse to introduce the game while teaching a class on automata about context-free languages; one student proved to be better at it than the rest of the class put together. She was the first educated person in her family and was a voracious reader who could spell many more words than she could pronounce. Naturally, she was less likely to be misled by tricky HIPEs; she knew words by their appearance.

How, indeed, do we "know" a word? Is it primarily by the word's meaning, by the word's sound, or by the word's appearance? Would you believe none of the above? My psychologist friends tell me that *the kinetic sense of how to produce a word* dominates, at least in some sense, the other aspects. Now, if that's really the case, then people who can't hear well, *especially* those who customarily communicate in a sign language, might be expected to have an advantage in playing HIPE. Their concept of a word might be expected to be short on sound and production (with the mouth) and thus relatively long on spelling.

Resolved to test this hypothesis, I introduced HIPE to a group of hearing-impaired employees at a government agency, who sat together and conversed in ASL daily at lunch. They found the game completely trivial; as fast as I wrote HIPEs on napkins, they wrote solutions around them. To them it was a mystery why anyone would think HIPE was any kind of challenge.

Looking back on this episode, it occurs to me that the fact that some of the folks at this table were professional cryptanalysts should have made me a bit hesitant about drawing conclusions. But is the connection between hearing loss and code-breaking a coincidence? My advice is, if you need someone to (say) read a garbled telegram, try your hearing-impaired friends first.

Here's a longer list of HIPEs, with no solutions provided. Of course, you can find solutions for any of them easily on your computer, by downloading a word list from the web and searching for a given HIPE combination by your favorite method. But I suggest trying out your brain first.

BG	CM	FC	FW	GC	GJ
KC	GD	GZ	HK	IJ	KG
KJ	LJ	LQ	MD	MQ	PJ
TJ	TK	TV	UH	UQ	XF
XG	XN	XQ	ZB	ZK	ZM
ZP	ZW				

AFY	AIE	AIQ	AKN	AKT
AMT	ANW	AOH	AOT	APK
ATG	AWG	BFR	BOF	BOJ
BPL	BPO	BTF	CAQ	CEC
CEK	CHG	CTY	CYH	CYO
CZE	DDB	DDM	DEK	DEQ
DHP	DSC	DSK	DTE	DWR
DYB	DYG	DYM	EGW	EKD
EOI	ETD	EWG	EZA	FEG
FEK	FSA	FSI	FTB	FTT
GEC	GEF	GGN	GGP	GHG
GNP	GOC	GSK	HDI	HDU
HLR	HMM	HPL	HPR	HSH
HSK	HTC	HTG	HTM	HYN
IAI	IAU	IDP	IEI	IFA
IJI	IMF	IOA	IOE	IOI
IOV	IPK	IUT	IXF	IZU
KAC	KBL	KCL	KEB	KSG
KSK	KSU	KTR	KUS	KYC
KYR	LAL	LCL	LDB	LDT
LEK	LEQ	LFL	LIH	LKH
LML	LNO	LPF	LPL	LSC
LTC	LTP	LYD	LYF	LYV

MBB	MBN	MBP	MCH	MEU
MKH	MND	MNL	MPF	MPG
MPM	MPP	MSK	MSM	MSP
MSU	MVE	MWI	NDD	NDJ
NDK	NGN	NKG	NKM	NNK
NYH	NYP	OAB	OAU	OAV
OEQ	OEU	OHO	OHY	OIE
OKC	OUA	OUQ	OWU	OYH
OYO	OYR	PEV	PIF	PIU
PFR	PPH	PSF	PSM	PSP
PYC	PYW	RDP	RDV	RFD
RJU	RLH	RMC	RMP	RNH
RPM	RSB	RTG	RYD	SDR
SHH	SIQ	SKR	SSK	SUO
SYW	TBL	THC	THT	TIW
TMU	TOZ	TYD	UAH	UDB
UEO	UFA	UMC	UXU	VEF
VEP	VEW	VYH	VYS	WAW
WFE	WNC	WNM	WNP	WNU
WSB	WSM	WSW	XAD	XIB
XTB	XYE	XYM	XYS	YAG
YFR	YHA	YOE	YRD	YSC
YSL	YSY	YUN	YZY	ZEF

CEBE	DROB	ECIB	ELEL
ERYO	EWHA	FELE	FRAR
GHAG	GUAG	FODI	HHOO
HTEE	KADA	LECU	LESL
LSEL	IKEA	ITCA	MECA
MEON	NELT	MYRO	NIKE
OELA	OFOL	OFTO	OOMM
OORK	OMUC	OWAD	OWNO
PIOC	PLEL	PONR	RESK
ROOR	SPES	TYPU	UBBU
UGUG	ZIPA		

Promised solutions to earlier HIPEs:

- gloWWorm, poWWow

- saVVy, reVVing

- hitcHHike, batHHouse

- sovereiGNTy

- bankruPTCy

- tooTHAChe

- chrysantHEMUm

- raSPBerry

- extRAORdinary

- coXSwain

- heaDQuarters

- witcHCRaft

- fraUDUlent

- eYEBrow, eYEBall

- larYNGitis

- swasTIKa, baTIK

- saXOPhone

- unBEKnown

- stomACHACHe

- insTANTANeous

I hope you've enjoyed HIPE[7] and will use the game to drive your friends crazy. Now back to mathematical puzzles.

---

[7]For those who think games like this are good for nothing, picture your author as a high-school senior trying to get into Harvard. He is asked to write about himself. He is from Fair Lawn, NJ, and is good at math. He is competing with kids who were raised by monks in Mandalay, played solo bassoon for the Queen of England, and cloned a moose at age 12. What can he say? Hoping to sucker the admissions officers into playing HIPE, he writes a light essay called "The HIPE Story" in which he describes how he and his friends started a local craze. Four years later, he is a senior at Harvard and overhears a tutor who had served on the admissions committee torturing a colleague with HIPEs. What's more, the tutor is *calling them HIPEs*.

So, I figure HIPEs got me into Harvard, and I owe them this chapter.

Two Dimensions and Three

Mathematicians have long since regarded it as demean-
ing to work on problems related to elementary geometry
in two or three dimensions, in spite of the fact that it it
precisely this sort of mathematics which is of practical
value.
—Branko Grünbaum and G. C. Shephard,
Handbook of Applicable Mathematics

For many of us, the first exposure to theorems and proofs is in high school,
studying Euclidean plane geometry. But the puzzles you are about to tackle
are far, far from Euclid's *Elements* and will test your appreciation of the two-
and three-dimensional world.

Coins on the Table

One hundred identical coins lie on a rectangular table, in such a way that no
more can be added without overlapping. (We allow a coin to extend over the
edge, as long as its center is on the table.)

Prove that you can start all over again and cover the whole table with 400
of these coins! (This time we allow overlap *and* overhang).

Comment: Implicitly assumed is that each coin is a perfect, homogeneous disc
of uniform but negligible thickness.

Four Points, Two Distances

Find all the ways to arrange four points on the plane so that they determine
only two different distances.

The Prisoner and the Dog

A woman is imprisoned in a large field surrounded by a circular fence. Outside the fence is a vicious guard dog that can run four times as fast as the woman, but is trained to stay near the fence. If the woman can contrive to get to an unguarded point on the fence, she can quickly scale the fence and escape. But can she get to a point on the fence ahead of the dog?

Tennis Mystery

The ball is "out" but no linesperson can call it. Where did it land?

Comment: In a major tennis tournament each linesman or lineswoman is responsible for one line, and calls "out" when a ball misses that line and lands on the wrong side of it. Annoyingly, there's a plausible "out" shot that this plan doesn't cover; where is it?

Double Cover by Lines

Using two full sets of parallel lines, you can cover the plane in such a way that every point belongs to exactly two lines. Can you do this in any other way, i.e., can you cover each point of the plane exactly twice using a set of lines containing lines in more than two different directions?

Comment: You could, for example, try taking all lines tangent to some fixed circle. This works great outside the circle but hits the points on the circle only once, and misses the inside entirely.

Time to ramp up to 3-space.

Curve on a Sphere

Prove that if a closed curve on the unit sphere has length less than 2π, then it is contained in some hemisphere.

Comment: This seems like it *ought* to be true, since the length of a great circle (the boundary of a hemisphere) is 2π. But how to prove it?

Laser Gun

You find yourself standing in a large rectangular room with mirrored walls. At another point in the room is your enemy, brandishing a laser gun. You and she are fixed points in the room; your only defense is that you may summon bodyguards (also points) and position them in the room to absorb the laser rays for you. How many bodyguards do you need to block all possible shots by the enemy?

Comment: "Infinitely many" is an acceptable answer, if it is right!

We conclude with a marvelous problem that actually has a connection to "real life." In many places the cost of posting or shipping a rectangular box is determined by *adding* the length, width, and height of the box, then looking up this figure in a table; obviously, the higher the figure, the greater the cost. Could it be possible to save money by packing a box into a larger but cheaper one?

Box in a Box

Let the cost of a rectangular box be given by the sum of its length, width, and height. Prove or disprove: It is impossible to fit a box into a cheaper box.

Comment: Clearly one can fit a long box into a shorter one, by making use of a diagonal; but it would seem that you'd have to give up too much on the other two dimensions. In *two* dimensions, that is, with rectangles, one can easily see using the triangle inequality that an analogous money-saving packing is impossible with rectangles. But the method seems to break down in 3-space.

Sources and Solutions

Coins on the Table

This nice puzzle came to me by way of computer scientist Guy Kindler, during a marvelous visiting year by each of us at the Institute for Advanced Study in Princeton.

Let us observe first that if we double the radius (say, from $1''$ to $2''$) of each of the original coins—as in Figures 7 and 8—the result will be to cover the whole table. Why? Well, if a point P isn't covered, it must be $2''$ or more from any coin center, thus a (small) coin placed with its center at P would have fit into the original configuration.

Now, if we could replace each big coin by four small ones that cover the same area, we'd be done—but we can't.

But, rectangles *do* have the property that they can be partitioned into four copies of themselves. So, let us shrink the whole picture (of big coins covering

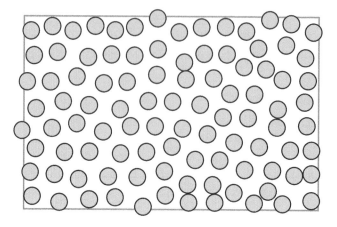

Figure 7. No more coins will fit without overlap.

the table) by a factor of two in each dimension, and use four copies (as in Figure 9) of the new picture to cover the original table!

Surprisingly (perhaps), this lovely but seemingly rather crude argument gives the best possible factor: replace the factor 4 by anything smaller, say 3.99, and the statement of the puzzle is no longer true.

To see this, we consider the limiting case where the table is very large and the coins numerous, so that boundary effects are negligible. Replace the

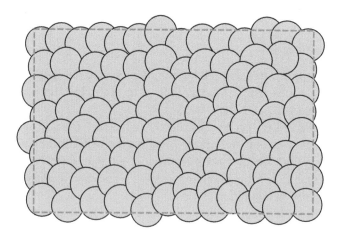

Figure 8. After doubling the table is covered.

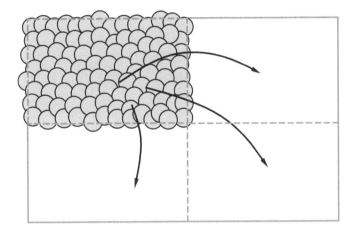

Figure 9. Four shrunken big-coin pictures cover the table.

table by a honeycomb-pattern bathroom floor with regular hexagonal tiles of diameter (say) 2. Since each tile could then partitioned into six equilateral triangles of side 1 and thus area $\sqrt{3}/4$, the tile itself has area $6 \times \sqrt{3}/4 = 3\sqrt{3}/2$.

We can cover the floor now by covering each tile with a coin whose boundary is the tile's circumscribing circle (see Figure 10).

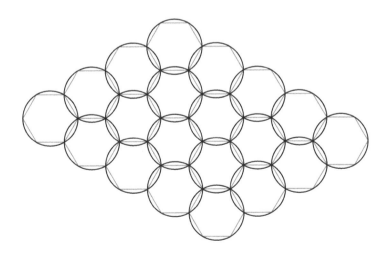

Figure 10. Covering the floor using a honeycomb pattern.

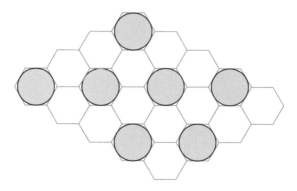

Figure 11. Coins slightly bigger than the inscribed circle.

Each coin then has radius 1 and thus area π. If the floor has area A then, ignoring boundary effects, the total area of the coins will be $\pi A/(\sqrt{3}/4) \sim 1.2092 \times A$.

Now, how thinly can we cover the floor without being able to add a non-overlapping coin? Let us use the same tiling, but this time we cover only every third tile (see Figure 11); and we cover it not with a circumscribed circle, but a coin placed over the center of the hexagon which is just a teensy bit bigger than the *inscribed* circle. This just barely prevents us from adding any more coins; how much is the coin area now?

Well, the coin radius is a tiny bit bigger than the altitude of one of the six equilateral triangles making up a hexagon—namely, $\sqrt{3}/2$. Hence, the coin area just exceeds $\pi \times (\sqrt{3}/2)^2 = 3\pi/4$.

It follows that the total coin area on the floor is as close as we like to $(1/3) \times (3\pi/4) \times A/(3\sqrt{3}/2) = \pi A/(6\sqrt{3}) \sim 0.3023 \times A$, one fourth of what we had before!

The result of all this is that we have proved not only the statement of the puzzle, but two not-so-easy extremal properties of disks in the plane. The first says that there is no better way to cover the plane by unit disks than by circumscribing the tiles in a hexagonal tiling, as we did above; the second says that there is no more efficient way to *prevent* the addition of a non-overlapping unit disk than by centering on every third hexagonal tile a disk slightly bigger than the inscribed disk, again as we did above.

If you think these properties are obvious to begin with, consider the even more obvious fact that the densest way to *pack* unit disks in the plane is to use an inscribed disk in each hexagon. This was not proved until the great Hungarian geometer László Fejes Tóth (1915–2005) did it in 1972!

Four Points, Two Distances

This wonderful lunch-table puzzle appeared as Problem 3a (submitted by S. J. Einhown and I. J. Schoenberg) in the "Puzzle Section" of the *Pi Mu Epsilon Journal* [16] in 1985. Later it showed up on page 1 of Nob Yoshigahara's *Puzzles 101* [61], where it was attributed to Dick Hess.

I have observed that very few people are able to come up with all six configurations; it seems that almost everyone has a mental block, or a reasoning flaw, which results in leaving one configuration out. Exactly *which* one gets left out is unpredictable; one challengee managed not to think of the square!

Anyway, here they are in Figure 12. The last of those pictured (the trapezoid) consists of four of the five vertices of a regular pentagon.

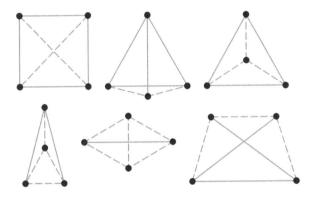

Figure 12. The six ways.

The Prisoner and the Dog

This nice escape problem was brought to my attention by Giulio Genovese and appears in Martin Gardner's *Mathematical Carnival* [24].

Define a "unit" as the radius of the field. If the prisoner were constrained to a smaller, concentric circle of radius r, where $r < \frac{1}{4}$, she would be able to maneuver herself to the farthest available point from the dog (see Figure 13); this is because the circumference of the small circle would be less than $\frac{1}{4}$ of the circumference of the field. But if r is close enough to $\frac{1}{4}$, the woman can then make a run for it straight to the fence. Her distance is only a hair over $\frac{3}{4}$ of a unit, but the dog has to go half way around the field, a distance of π units. Since $\pi > 3$, this is more than four times further than the woman has to run.

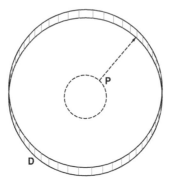

Figure 13. The point from which the prisoner (P) sprints for the fence.

The dog's factor of 4 speed advantage can be stretched to 4.6033388, at which point best strategy by both parties will result in a dead heat. For more details, see the May 2001 edition of IBM's puzzle site "Ponder This."[8]

Tennis Mystery

It was puzzle maven and tennis enthusiast Dick Hess who pointed out this little glitch to me. Figure 14 shows the ball's mark; it is a service fault, since it falls clean out of the intended service box, yet it is neither "long" nor "wide." It doesn't help if there's a machine calling the back service line. One wonders if this shot is often miscalled.

Figure 14. Who is supposed to call this service fault?

---

[8]http://domino.research.ibm.com/Comm/wwwr_ponder.nsf/solutions/May2001.html.

Double Cover by Lines

This one will disappoint some readers—the answer is yes (assuming the Axiom of Choice); in fact, there are infinitely many ways to do it. But the proof requires transfinite induction (!) and does not leave you with any geometry you can wrap your mind around. The problem (and its solution) were provided to me by physicist Senya Shlosman, who is unaware of its origin.

Nonetheless, I like this solution as an example of an easy application of a powerful tool. The idea is this: we start off with three lines that cross one another, so that we already have our three directions. Now, let κ be the least ordinal of the cardinality of the continuum (the number of points on a line, points in the plane, or angles in the plane). We induce on the set of ordinals below κ. Each of these is either a successor ordinal (like 17, 188, or $\omega + 1$) or a limit ordinal (like ω, the first infinite ordinal); and each has cardinality strictly *less* than the cardinality of the continuum. The *number* of ordinals below κ is the continuum, so we may label the points of the plane by these ordinals. The points now form a "well-ordered" set, meaning that every nonempty subset of the plane has a point with smallest label.

Now to do transfinite induction, we imagine that we are at stage σ and have constructed a line for each ordinal less than σ, no point being covered more than twice. We now take the smallest-labeled point of the plane that is not currently double-covered and add a line through that point. How do we know we can do that without triple-covering some other point? Well, the number of lines so far in our set is only the cardinality of σ, thus the number of *pairs* of such lines is less than the cardinality of the continuum; therefore there is an angle we can pick for *this* line that misses all the points which are currently double-covered. Voila!

Does this seem like cheating? It probably should; no way you can do this construction in any useful manner. Does it mean there *isn't* some nice way to double-cover the plane by lines? No, but I haven't found one; neither has Shlosman.

Curve on a Sphere

I got this puzzle from physicist Senya Shlosman, who heard it from Alex Krasnoshel'skii. The solution proposed by Senya is as follows.

Pick any point P on the curve, travel half way around the curve to the point Q, and let N (standing for North Pole) be the point half-way between P and Q. (Since the distance $d(P, Q)$ from P to Q is less than π, N is uniquely defined). N determines an "equator" and if the curve lies entirely in the northern hemisphere, we are done. Otherwise, the curve crosses the equator, and let E be one of the points at which it does so. Then, we observe that $d(E, P) + d(E, Q) = \pi$, since if you poke P through the equatorial plane to P' on the other side, P' is antipodal to Q; hence, $d(E, P') + d(E, Q) = \pi$.

However, for any point X on the curve, $d(P, X) + d(X, Q)$ must be less than π, and this provides the desired contradiction.

Omer Angel, of the University of British Columbia, came up with a totally different sort of proof that is less elementary, but still elegant and educational. Let C be our closed curve and $\hat{C}$ its convex closure, that is, the smallest convex set containing C. If C is not contained in a hemisphere, then $\hat{C}$ contains the origin 0; otherwise, there would be a plane through 0 cutting off $\hat{C}$ from 0. Thus, by Carathéodory's Theorem (see below), there is a set of four points on C some convex combination of which is 0. Putting it another way, the tetrahedron whose vertices are those four points contains the origin.

Now, move the points continuously toward one another along the curve. When the points merge their tetrahedron will no longer contain the origin, so somewhere during the process, there was a time when the origin lay on a *face* of the tetrahedron. The three points that determine that face lie on a great circle, and each pair has a shortest route along that circle not containing the third point. Hence, the sum of the pairwise distances of the three points is 2π, impossible since they all lie on C.

The mathematician Constantin Carathéodory (1873–1950) proved many elegant theorems, of which one of the best known states that if v is in the convex closure of certain points in d-space then v is already in the convex closure of a subset of at most $d + 1$ of them.

To prove this, we observe that being in the convex closure of a set is equivalent to being expressible as a finite linear combination of points of the set with positive coefficients which sum to 1. Let $k > d + 1$, and put $v = \sum_{i=1}^{k} a_i v_i$ where $\sum_{1}^{k} a_i = 1$ and each $a_i > 1$.

Since there are more than d of them, the vectors $v_1 - v_i$, $i = 2, \ldots, k$ are linearly dependent; thus, there are b_i's not all zero such that $\sum_2 b_i(v_1 - v_i) = 0$. Put $b_1 := -\sum_{j=2}^{k} b_j$; then $\sum_{i=1}^{k} b_i v_i = 0$ and $\sum_{i=1}^{k} b_i = 0$ but some $b_j \neq 0$, thus at least one b_i is positive.

Thus, for any real number r, $v = \sum a_i v_i - r \sum b_i v_i = \sum (a_i - rb_i)v_i$. In particular, let r be the smallest ratio a_i/b_i (achieved, say, at $i = j$) for $b_i > 0$. Then, r is positive and $a_i - rb_i \geq 0$ for all i; thus, we have v represented as a convex combination at least one of whose coefficients (namely, $a_j - rb_j$) is zero, so we've got v in the convex closure of at most $k - 1$ points. Repeat until you are down to $d + 1$ points.

Laser Gun

This puzzle was brought to my attention by Giulio Genovese, who got it from Enrico Le Donne; they traced it to a 1990 Mathematics Olympiad in St. Petersburg [17]. Amazingly, sixteen bodyguards will do the trick!

It seems that the puzzle has set off a chain reaction of research on the "security problem," namely, the question of which room shapes other than the rectangle have the property that finitely many bodyguards suffice. The question has not been fully answered even for polygons with rational angles, but it does follow from the work of Eugene Gutkin [34] that the only *regular* secure polygons are the equilateral triangle, the square, and the regular hexagon.

Returning to the puzzle at hand, the basic idea of the proof is this. View the room as a rectangle in the plane, with you at P and your enemy at Q. You can now tile the plane with copies of the room, by repeatedly reflecting the room about its walls, each copy containing a new copy of your enemy (see Figure 15).

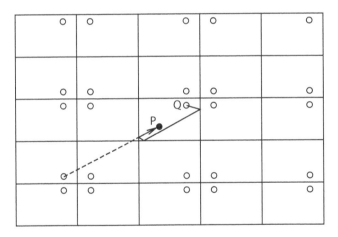

Figure 15. A tiling of the plane by reflected copies of the room (and your enemy).

Every possible shot by the enemy can be represented on this picture by a straight line from *some* copy of Q to P; every time such a line crosses a boundary between rectangles, the real laser beam would be bouncing off a wall. In the figure, one such (dotted) line is indicated; a solid line shows the corresponding path of the laser beam back in the original room.

Your object is going to be to intercept every shot at its halfway point. To do this, you first "trace" a copy of the plane tiling in Figure 15, then nail it to the plane at your position P, and finally shrink the copy by a factor of two

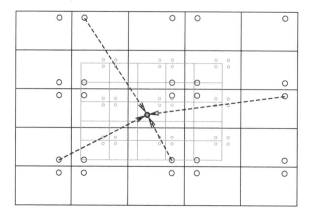

Figure 16. A shrunken copy of the plane is superimposed with P on top itself.

vertically and horizontally. The many copies of Q on the shrunken copy will be our bodyguard positions; they serve our purpose because each copy of Q on the original tiling appears halfway between it and you on the shrunken copy.

In Figure 16, the shrunken copy is in gray, and some virtual laser paths are indicated; you can see that they pass, at their half-way points, through the corresponding smaller dots in the gray grid.

Of course, there are infinitely many such points, but we claim that they are all reflections of the right set of 16 points in the original room. Four of the points will already be in the original room. The four points in the room to the left of the original room can be relected back to give four new points, and similarly for the room above the original one. Finally, the four points in the room above *and* to the left of the original room can be reflected twice to provide the last four points in the original room. In Figure 17, the twelve new points (centers filled in gray) have been added in black to the original rectangle. A virtual laser path is added, with its corresponding real path crossing one of the new points.

Since every room looks exactly like the original or one of the other three we just examined, all the bodyguard points in the plane are reflections of the sixteen points that we have now identified in the original room. Since every line from a copy of Q passed through a reflected bodyguard, the actual shot hits a "real" bodyguard at its halfway point (if not sooner) and is absorbed.

If the locations of P and Q are carefully chosen, some of the bodyguard locations will coincide; but in general the full sixteen are needed.

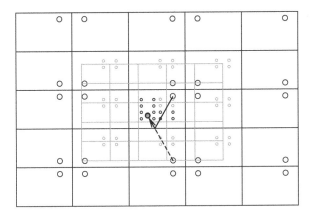

Figure 17. New points in the original rectangle account for all the bodyguard positions.

Box in a Box

This lovely puzzle was communicated to me by Anthony Quas (University of Victoria) who heard it, and the solution below, from Isaac Kornfeld, a professor at Northwestern University. Kornfeld had heard the puzzle many years ago in Moscow.

Let B_ε be the expansion, by an amount $\varepsilon > 0$, of a box B; that is, the set of all points in space within distance ε of some point of B. If B is $a \times b \times c$ then B_ε will be approximately $(a + 2\varepsilon) \times (b + 2\varepsilon) \times (c + 2\varepsilon)$, but with rounded edges and corners. The precise volume of B_ε will be abc (the volume of B) plus $2abe + 2ace + 2bce$ (the volume of the slabs added to the six sides) plus $4a\pi\varepsilon^2/4 + 4b\pi\varepsilon^2/4 + 4c\pi\varepsilon^2/4$ (the volume of the 12 "molding strips" added to the edges—each having a quarter-circle cross section—plus $4\pi\varepsilon^3/3$, since the eight knobs added to the corners add up to a sphere. Altogether,

$$\mathrm{Vol}(B_\varepsilon) = \frac{4}{3}\pi\varepsilon^3 + (a + b + c)\pi\varepsilon^2 + 2(ab + ac + bc)\varepsilon + abc\,.$$

It's difficult to illustrate B_ε accurately so instead we drop to the plane, and show in Figure 18 what B_ε would look like if B were just an $a \times b$ rectangle. Here the formula for the area of the expanded figure would be

$$\mathrm{Area}(B_\varepsilon) = \pi\varepsilon^2 + 2(a + b)\varepsilon + ab\,.$$

Returning to three dimensions, if box A (with dimensions, say, a', b', and c') lies inside box B, then A_ε lies inside B_ε, for any $\varepsilon > 0$. Hence $\mathrm{Vol}(A_\varepsilon) < \mathrm{Vol}(B_\varepsilon)$. But, if we take ε to be *huge*, the dominant term in the

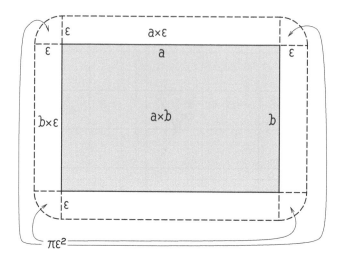

Figure 18. An ε-expanded $a \times b$ rectangle.

difference of their volumes is

$$(a + b + c)\pi\varepsilon^2 - (a' + b' + c')\pi\varepsilon^2.$$

Since this term must be nonnegative, B is the costlier box.

The problem also appeared in Tournament of the Towns, as Problem 5, in the 1998 Fall Round Senior Paper, Advanced Level. The solution given there was a different one, contributed by Andrei Storozhev, an expatriate Russian now working for the Australian Mathematics Trust. Storozhev's solution relies on the observation that the surface area of the inside box (Box A) must be smaller than that of Box B. We can justify this statement by projecting each face of Box A outward, perpendicular to itself, so that it captures a piece of the surface of Box B. Since these six pieces are disjoint and each is at least as large as the corresponding face of Box A, the statement follows.

We can represent this area comparison algebraically by $2a'b' + 2b'c' + 2c'a' < 2ab + 2bc + 2ca$, but we also know that $a'^2 + b'^2 + c'^2 < a^2 + b^2 + c^2$ by comparing the *diagonals* of the two boxes. Adding these two inequalities gives $(a' + b' + c')^2 < (a + b + c)^2$, and we are done!

Lines and Graphs

The human heart likes a little disorder in its geometry.

—Louis de Bernieres (1954–)

Now we're down to one dimension—to lines, which you know about, and graphs, which maybe you don't. A graph is just a collection of points, called *vertices*, some pairs of which constitute *edges*. Often the vertices of a graph are realized by points in the plane, and its edges by line segments or curves connecting one vertex to the other. If this can be done in such a way that the curves do not cross one another, the graph is said to be *planar*.

Bracing the Grid

Suppose that you are given an $n \times n$ grid of unit-length rods, jointed at their ends. You may brace some subset S of the small squares with diagonal segments (of length $\sqrt{2}$).

Which choices of S suffice to make the grid rigid in the plane?

Figure 19 shows an (apparently) inadequately braced 3×3 grid.

Figure 19. A grid with some—but not enough—squares braced.

Touring an Island

Aloysius is lost while driving his Porsche on an island in which every intersection is a meeting of three (two-way) streets. He decides to adopt the following algorithm: starting in an arbitrary direction from his current intersection, he turns right at the next intersection, then left at the next, then right, then left, and so forth.

Prove that Aloysius must return eventually to the intersection at which he began this procedure.

Comment: A graph in which every vertex is incident to three edges is said to be "cubic." In this problem there is a consistent notion of "right" or "left," which is always the case when the vertices of a graph lie on a plane, with curves (streets) representing edges. On Aloysius' island it is not actually necessary for the graph to be *planar;* we can allow places (bridges or tunnels) where edges cross.

Wires under the Hudson

Fifty identical wires run through a tunnel under the Hudson River, but they all look the same, and you need to determine which pairs of wire-ends belong to the same wire. To do this you can tie pairs of wires together at the west end of the tunnel and test pairs of wire-ends at the east end to see if they close a circuit; in other words, you can determine whether two wires are tied together at the other end.

How many trips across the Hudson do you need to accomplish your task?

Bugs on Four Lines

You are given four lines in a plane in general position (no two parallel, no three intersecting in a common point). On each line a ghost bug crawls at some constant velocity (possibly different for each bug). Being ghosts, if two bugs happen to cross paths they just continue crawling through each other uninterrupted.

Suppose that five of the possible six meetings actually happen. Prove that the sixth does as well.

While we're on the subject of bugs...

Spiders on a Cube

Three spiders are trying to catch an ant. All are constrained to the edges of a cube. Each spider can move at least one third as fast as the ant can. Prove that the spiders can catch the ant.

The next puzzle brings us to a lovely general theorem of graph theory.

Impressionable Thinkers

The citizens of Floptown meet each week to talk about town politics, and in particular whether or not to support the building of a new shopping mall downtown. During the meetings each citizen talks to his friends—of whom there are always an odd number, for some reason—and the next day, changes (if necessary) his opinion regarding the mall so as to conform to the opinion of the majority of his friends.

Prove that eventually, the opinions held every *other* week will be the same.

Comment: Since there are only a finite number of opinion-patterns (2^n, if there are n citizens), it is clear that after a while the pattern must cycle. The claim here is that the cycle has period only 2 (or 1). Why on earth should that be true?

We conclude with a creature who wants to stay on his graph.

Lemming on a Chessboard

On each square in an $n \times n$ chessboard is an arrow pointing to one of its eight neighbors (or off the board, if it's an edge square). However, arrows in neighboring squares (diagonal neighbors included) may not differ in direction by more than 45 degrees.

A lemming begins in the center square, following the arrows from square to square. Is he doomed to fall off the board?

Sources and Solutions

Bracing the Grid

This intriguing (and potentially practical) puzzle was passed to me by master geometer Bob Connelly of Cornell, based on the work of Ethan Bolker and Henry Crapo [8].

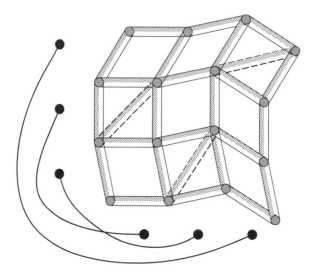

Figure 20. Our floppy grid and its corresponding row-and-column graph.

It helps to turn this into a graph-theoretical problem, but not in the most obvious way (with vertices as joints and edges as rods). Instead, suppose that you have put in your braces; now envision a graph G whose vertices correspond to the rows of squares and the columns of squares. Each edge of G corresponds to a row and column whose intersecting square is braced, so that the number of edges of G is the same as the number of braces you used.

The grid pictured in Figure 19 is shown again, but with its graph, in Figure 20.

If a row and a column are adjacent in G, the vertical rods in the row are all forced to be perpendicular to the horizontal rods in the column. If G is a *connected* graph, meaning that there is a path from any vertex to any other, then *all* the horizontal rods must be perpendicular to all the vertical ones. Thus all the horizontal rods are parallel to one another, and similarly for the vertical ones, and now it is clear that the grid is rigid.

On the other hand, suppose that the graph is disconnected and let C be a *component*, that is, a connected piece of G which has no edges to the rest of G. Then, there is nothing to prevent any vertical rod in a C-row or any horizontal rod in a C-column from flexing relative to the other rods in the grid.

Thus, the criterion for rigidity is exactly that the graph G be connected. Since G has $2n$ vertices, it must have at least $2n - 1$ edges in order to be connected (you can prove this easily by induction if you haven't seen it before),

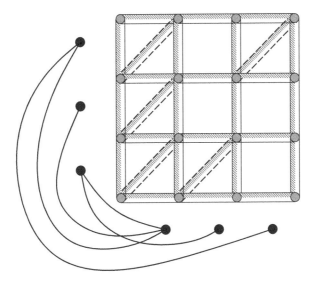

Figure 21. An adequately braced grid and its corresponding connected graph.

hence you must have at least $2n - 1$ braces to make the grid rigid. Notice, however, that they cannot be "just anywhere."

Figure 21 shows an efficiently braced 3×3 grid and its corresponding graph. For practice, you might want to compute the total number of ways to rigidify the 3×3 grid with the minimum number (five) of braces. A theorem of graph theory (to the effect that every connected graph has a connected subgraph, called a "spanning tree," with the minimum number of edges) tells us that if more than $2n - 1$ squares are braced *and the grid is rigid*, then there are ways to remove all but $2n - 1$ of the braces and preserve rigidity.

Touring an Island

A version of this puzzle was taken for the CMU webpage "The Puzzle Toad," mentioned earlier, from the book by G. A. Galperin and A. K. Tolpygo, *Moscow Mathematical Olympiads* [23].

Between intersections Aloysius' current state can be characterized by a triple consisting of the edge he's on, the direction he's going on the edge, and the type of his last turn (right or left). Since there are only finitely many such triples, there will be a first time when Aloysius hits a triple twice; this can only happen at his starting edge!

Wires under the Hudson

This is a variation of a puzzle publicized by Martin Gardner, sometimes called the Graham-Knowlton Problem. To electricians it is the "WIP" (wire identification problem). In Gardner's version, you can tie any number of wires together at either end, and test at either end, as well. The following solution is found in *Recreation in Mathematics* by Roland Sprague [54] and also in a recent paper by three young computer scientists, Navin Goyal, Sachin Lodha, and S. Muthukrishnan [33]; it satisfies our additional constraints and involves only two operations at each end (thus three river crossings, not counting additional crossings to untie and perhaps actually *use* the wires). The solution is not unique, however, so even if your three-crossing solution is different, it may be just as good.

Suppose that the wires are labeled $w_1, w_2, \ldots, w_n$ at the west end of the tunnel and $e_1, \ldots, e_n$ at the east end. On our first visit to the west end we tie w_1 and w_2 together, w_3 and w_4, w_5 and w_6, etc. until all except w_{49} and w_{50} are paired.

We then test pairs of wire-ends at the east end of the tunnel until we have discovered all the tied pairs. For example, we may find that e_4 and e_{29} are tied, e_2 and e_{15} are tied, e_8 and e_{31} are tied, and so forth, and finally that e_{12} and e_{40} are bachelors.

Next we return to the west end, untie all pairs, and instead tie w_2 to w_3, w_4 to w_5, etc. until all but w_1 and w_{50} are paired.

Finally we test pairs at the east end until, as before, we have identified all the paired wire-ends. To continue the example, the new pairs might include e_{12} and e_{15}, e_{29} and e_2, and e_4 and e_{31}, with e_{40} and e_8 as bachelors.

Amazingly, this simple procedure suffices to identify all the wires!

Observe that the one east wire-end which was paired the first time but not the second (in our example, this is e_8) must belong to w_1. The east wire-end with which e_8 was paired the first time (here, e_{31}) must therefore belong to w_2. But then, w_3 must belong to the east wire-end to which e_{31} was paired the *second* time, namely e_4. Proceeding in this fashion, we find that w_4 belongs to e_{29} (e_4's mate on the first round), w_5 belongs to e_2 (e_{29}'s mate on the second round), and so forth. Eventually the sequence will end with w_{50} belonging to e_{40}.

If the number of wires (say, n) had been odd, we'd have left only w_n out of the pairings the first time, and w_1 the second time; the rest works pretty much the same way.

Bugs on Four Lines

This puzzle was passed to me by Matt Baker, of Georgia Tech. It is sometimes called "the four travellers' problem" and appears also on the website "Interactive Mathematics Miscellany and Puzzles" at http://www.cut-the-knot.org.

By far the most elegant solution I have seen requires that you lift the problem into space, by means of a time axis. Suppose that every pair of bugs meets except bug 3 with bug 4. Construct a time axis perpendicular to the plane of the bugs, and let g_i be the graph (this time, in the sense of graphing a function) of the ith bug in space. Since each bug travels at constant speed, each such graph is a straight line; its projection onto the plane of the bugs is the line on which that bug travels. If (and only if) two bugs meet, their graphs intersect.

The lines g_1, g_2, and g_3 are coplanar since all three pairs intersect, and the same applies to g_1, g_2 and g_4. Hence, all four graphs are coplanar. Now g_3 and g_4 are certainly not parallel, since their projections onto the original plane intersect, thus they do intersect on the new common plane. So bugs 3 and 4 meet as well.

Spiders on a Cube

This puzzle has the same provenance as Touring an Island above.

One simple way to catch the ant is to use each of two spiders to "control" an edge. To control an edge PQ, the spider chases the ant off the edge if necessary, then patrols the edge making sure he is at all times at most one

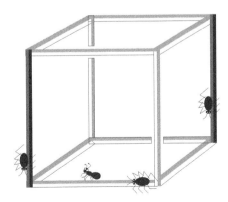

Figure 22. With the two black edges controlled, the ant is hunted down on the gray network.

third the distance from P (respectively, Q) that the ant is. This is possible since the distance from P to Q along the edges of the cube, if you are not allowed to use the edge PQ, is three times the length of the edge.

If we choose for our two controlled edges two *opposite* edges (other choices are equally good), we find that with these edges—including their endpoints—removed, the rest of the edge-network contains no cycles. (See Figure 22.) It follows that the third spider can simply chase the ant to the end of a patrolled edge, where she will meet her sad fate.

Impressionable Thinkers

This puzzle was suggested to me by Sasha Razborov, at the Institute for Advanced Study; he tells me that it was considered for an International Mathematics Olympiad, but rejected as too hard. It was posed and solved in a paper by E. Goles and J. Olivos [31].

To prove that the opinions eventually either become fixed or cycle every other week, think of each acquaintanceship between citizens as a pair of arrows, one in each direction. Let us say that an arrow is currently "sour" if the opinion of the citizen at its tail differs from *next week's* opinion of the citizen at its head.

Consider the arrows pointing out from citizen Clyde at week $t - 1$, during which (say) Clyde favors the mall. Suppose that m of these are sour. If Clyde still (or again) favors the mall at week $t + 1$, then the number n of sour arrows pointing *toward* Clyde at week t will be exactly m.

If, however, Clyde is against the mall at week $t + 1$, n will be strictly less than m since it must have been that the majority of his friends opposed the mall at week t. Therefore a majority of the arrows out of Clyde were sour at week $t - 1$ and now only a minority of the arrows into Clyde at week t are sour.

The same observations hold, of course, if Clyde is against the mall at week $t - 1$.

But, here's the thing: *every* arrow is out of *someone* at week $t - 1$, and into someone at week t. Thus, the total number of sour arrows cannot rise between weeks $t - 1$ and t and, in fact, will go strictly down unless every citizen had the same opinion in week $t - 1$ as in week $t + 1$.

But, of course, the total number of sour arrows in a given week cannot go down forever and must eventually reach some number k from which it never descends. At that point, every citizen will either keep his opinion forever or flop back and forth every week.

The puzzle can be generalized considerably, for example, by adding weights to vertices (meaning that some citizens' opinions are more highly prized than others'), allowing loops (citizens who consider their own current opinions as well), allowing tie-breaking mechanisms, and even allowing different thresholds for pro-mall and anti-mall opinion-changes.

Lemming on a Chessboard

This delightful puzzle was invented by Kevin Purbhoo when he was a high-school student at Northern Secondary School in Toronto. Purbhoo has since gotten a PhD in mathematics from UC Berkeley and is working on something called "tropical geometry" as a postdoc at the University of British Columbia. The puzzle was communicated to me by Ravi Vakil, of Stanford.

The lemming is indeed doomed. One way to see this (spotted independently by Vakil and me) is to imagine that the lemming can move to *any* neighboring square, but must turn to face in the direction of the arrow found there. Then the lemming cannot turn 360 degrees around, because if he could you could shrink the cycle on which he does that until it collapses into a contradiction. But the real lemming, if he is to stay on the board, must eventually cycle, and when he does so, he will have to make that 360° turn.

Purbhoo's own solution, from high school days, employs induction. If the lemming stays on the board, he must, as we have already noted, eventually settle into a cycle. Let C be the smallest-area cycle (on any board) on which this can happen, and suppose it's a clockwise cycle. Cut the whole board down to C and its interior, then rotate all the arrows 45° clockwise to force a smaller cycle!

Games and Strategies

If life doesn't offer a game worth playing, then invent a new one.

—Anthony J. D'Angelo, *The College Blue Book*

If we didn't play games we'd need to invent them anyway, as many problems in mathematics are best viewed as games. In fact, you've already seen some puzzles that involve computing the best strategy for some game; here are some more.

We begin with a simple question about a game people actually do play—poker.

Poker Quickie

What is the best full house?

Comment: You may assume that you are playing straight five-card stud poker (everyone gets five cards, all closed, no exchange) with (say) five companions, using a single normal deck. As a result of God owing you a favor, you are entitled to a full house, and you get to choose the full house you will get. Which should you choose?

Recovering the Polynomial

The Oracle at Delphi has in mind a certain polynomial (in the variable x, say) with nonnegative integer coefficients. You may query the Oracle with any integer x, and the Oracle will tell you the value of $p(x)$.

How many queries do you have to make to determine p?

A Game of Desperation

On a piece of paper is a row of n empty boxes. Tristan and Isolde take turns, each writing an "S" or an "O" into a previously blank box. The winner is the one who completes an "SOS" in consecutive boxes. For which n does the second player (Isolde) have a winning strategy?

Urn Solitaire

Before you is an urn containing some green balls and some red ones (at least one of each). In Round 1 of this game, you draw a ball blindly and note its color. You then continue to draw balls (always randomly) until you get one of the *other* color; that one is then returned to the urn.

Round 2 and successive rounds are repetitions of Round 1. You play until the urn is empty; if the last ball drawn is green, you win.

How many green balls and how many red balls should you start with in the urn to maximize your probability of winning?

Pirates and Gold

A pirate ship with a crew of 100 has captured a treasure chest containing some gold coins, and protocol demands that they be divvied up in the following manner. Each pirate, in rank order from the captain on down, will in turn make a proposal for who gets how many coins. All pirates including the proposer will vote on each proposal, tie going to the proposer. If a proposal is accepted, the coins are distributed accordingly and the process terminates; but if it is voted down, the proposer walks the plank and it's the turn of the next pirate in rank to try to do better.

You may assume that the pirates are cunning, greedy, and cautious, their highest priority being to avoid any possibility of having to walk the plank. If a pirate is indifferent between two actions, he is unpredictable.

How many coins must there be in the chest in order for the captain himself to guarantee his own survival?

Frames on a Chessboard

You have an ordinary 8×8 chessboard with red and black squares. A genie gives you two "magic frames," one 2×2 and one 3×3. When you place one of these frames neatly on the chessboard, the 4 or 9 squares they enclose instantly flip their colors.

Can you reach all 2^{64} possible color configurations?

More Frames on a Smaller Board

This time you have only a 6×6 board, and in each of the 36 squares is an integer. You can pick any 2×2, 3×3, 4×4, 5×5, or 6×6 subsquare and add 1 to each number inside the subsquare. Is it the case that, given any initial configuration, you can get to one in which every number is a multiple of 3?

Poker and many other games involve an element of bluffing, which can be a very complex phenomenon. For the next game, we strip to the bare bluffing essentials.

A Simple Bluff

Consider the following simple bluffing game. Louise and Jeremy ante one dollar each; Louise takes and examines a card from a shuffled deck. She may now raise the bet by \$10 (by adding her \$10 to the pot) or leave it as is. If she fails to raise, she wins the pot when her card is a spade and loses it otherwise.

If Louise raises, Jeremy may call (by adding his own \$10 to the pot) or fold. If Jeremy folds, Louise takes the money in the pot, thus winning Jeremy's ante. If Jeremy calls and Louise's card is a spade, Louise again takes the pot, this time with \$11 of Jeremy's money in it. But if her card is not a spade, Jeremy gets the pot instead.

Who has the advantage in this game? Would it make a difference if the bet size were changed to something other than \$10?

We conclude with an intriguing form of the classic game Nim.

Chinese Nim

On the table are two piles of beans. Alex must either take some beans from one pile or the same number of beans from each pile; then Beth has the same options. They continue alternating until one wins the game by taking the last bean.

What's the correct strategy for this game? For example, if Alex is faced with piles of size 12,000 and 20,000, what should he do? How about 12,000 and 19,000?

Sources and Solutions

Poker Quickie

This was sent to me by Stan Wagon, who found it in Aaron Friedland's book, *Puzzles in Math and Logic* [18].

The point is that, in effect, all full houses with three aces are equally high, because there can only be one such hand in a deal from a single deck. But there are *other* hands that can beat them: any four-of-a-kind, of which there are always 11 varieties possible, and more relevantly any straight flush. Since AAA99, AAA88, AAA77, AAA66, and AAA55 kill the most straight flushes (thirteen—each ace kills only one, but each spot card kills five) they are the best full houses. If you greedily insist on AAAKK there are $40 - 6 = 34$ possible straight flushes that can beat you—even more if you haven't got the four suits covered—instead of just $40 - 13 = 37$.

Recovering the Polynomial

I got this one from Joe Buhler (Reed College) who believes it may be quite old.

As you may already have figured out, it only takes two queries: the Oracle's answer to $x = 1$ (say, n) tells you that no coefficient can exceed n. Then, you can send in $x = n + 1$, and when you expand the Oracle's answer base $n + 1$, you have the polynomial!

Joe notes that without the restriction that you must give the Oracle an integer, you can win in one step with $x = \pi$. Of course, you'll need to trust that the Oracle will find a way to pass $p(\pi)$ to you in a finite amount of time; if she instead gives you a decimal expansion digit by digit, you won't know when to cut her off.

A Game of Desperation

The SOS game was brought to my attention by my PhD student, Rachel Esselstein; it is discussed, along with many other games, in Tom Ferguson's *Game Theory Text*, which you can find at http://www.math.ucla.edu/~tom/ Game_Theory/Contents.html. The version above appeared at the 28th Annual USA Mathematical Olympiad in 1999.

The game seems confusing and impenetrable until you realize that the only way you can force a win on your next move is to force your opponent to play into the configuration S-blank-blank-S (henceforth to be called a "pit"). Thus, for example, Tristan can win when $n = 7$ by placing an S in the middle, then another at the end farthest from Isolde's response, to make a pit. After a move by each player on the response end, Isolde must play into the pit and lose.

The same applies for any odd n greater than 7, as Tristan can play an S anywhere at least 4 spaces from the ends, then form a pit on one side or the other and wait.

When n is even Tristan has no chance, as there will never be a time when Isolde has only pits to play in; when she moves, there is always an odd number of blank squares to play in. Instead, when n is even and large, Isolde wins by playing an S far from the ends and from Tristan's first move. However, if Tristan begins with an O, Isolde cannot put an S next to it, so she needs extra room.

In the $n = 14$ case, if Tristan writes an O in place 7 (of 1 through 14), Isolde's best response is an S in position 11 (threatening to make a pit with an S in 14). Tristan can counter this with an O in 13 or 14 (or an S in 12 or 13), and now Isolde would like to make a pit with S in position 8 but can't, as Tristan would then win with S in 6.

Thus, $n = 14$ is a draw; Isolde needs n to be even and at least 16. To wrap up, Tristan wins when n is odd and at least 7, Isolde when n is even and at least 16; all other values of n lead to a draw with best play.

Urn Solitaire

This puzzle (presented slightly differently) appears as Problem 2.16 in Martin Gardner's *The Colossal Book of Short Puzzles and Problems* [30], but the solution is given without proof. Indeed, the proof presented in the paper [45] cited there is three pages long and too technical either for Gardner's book or this one.

As it happens, however, there is an easy way to see that the probability of winning Urn Solitaire is exactly $\frac{1}{2}$ irrespective of the red-green population of the urn. Below is the argument suggested by Sergiu Hart of Hebrew University, Jerusalem, who called the problem to my attention.

It sometimes pays, in analyzing a random process, to move the randomness to a different place. Here, it is useful—and permissable—to imagine that prior to each round of play, the remaining balls are lined up in random order and you choose them from the left. In that case, just before the last round, all remaining balls are the same color. Just before the *next-to-last* round, there are balls of both colors with all the red balls on the left and all the green balls on the right, or vice versa. Regardless of how many balls there are of each color at this point (or originally), these two orders are equally likely. Since the first results in a win and the second in a loss, your probability of winning is $\frac{1}{2}$.

Sergiu points out that Urn Solitaire ends up being in a sense isomorphic to "The Lost Boarding Pass" [59, page 35].

Pirates and Gold

I was reminded of this old teaser by Dartmouth graduate student Giulio Genovese. Like many games, it yields to retrograde analysis. Let's number the pirates in reverse rank order, and for now say the number of coins is n. If it comes down to lowly P_1, he of course takes all the gold; let's hope he can bring the ship to port by himself!

This is a moot point, though, because if it ever comes down to P_1 and P_2, P_2 will vote himself all the coins, then survive to command the ship.

P_3, if he ever gets to make a proposal, can assure himself of P_1's vote by offering him one coin, taking the remaining $n - 1$ for himself.

Hence P_4's best course would be to bribe P_2 with one coin; that's all it takes, since P_2 gets nothing if P_4's proposition fails.

P_5 needs two votes, and a coin to each of P_1 and P_3 will do the trick.

By now the pattern is beginning to emerge, and we have a statement that we can try to prove by induction: If there are enough coins to do this, and an odd number of pirates remain, the proposer should offer one coin to each remaining odd-ranked pirate; if an even number of pirates remain, the proposer should offer one coin to each remaining even-ranked pirate. The induction hypothesis specifies not only that these proposals are correct, but also that all the coin-receiving pirates will vote "aye." The proof now is straightforward.

What of the captain's fate? He'll need 49 coins to offer to the even-ranked pirates below 100 if he wants to guarantee survival.

Frames on a Chessboard

This puzzle was suggested by Ehud Friedgut of Hebrew University; it is a variation of one that appeared on an Israeli youth mathematics contest. In the contest, the frames were 3×3 and 4×4, and a counting argument tells you that you cannot reach every color configuration. The point is that the order in which the frames are laid down is irrelevant; all you need to know is which of the 5^2 ways to put the 4×4 frame down, and which of the 6^2 ways to put the 3×3 frame down, are utilized. Altogether there are thus $2^{25} \times 2^{36} = 2^{61}$ ways to *try* to get color configurations; not enough.

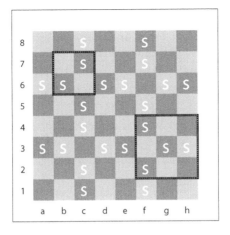

Figure 23. Special squares (each indicated with a white "S") and sample frames.

With Friedgut's modification, however, there are $2^{49} \times 2^{36}$ ways to select frame locations, enough in theory to get all 2^{64} color configurations. But can you?

Label squares as "special" if they are in row 3 or 6 or column c or f but not both (see Figure 23); then, every 2×2 or 3×3 frame covers an even number of special squares.

Since the board starts with an even number of black special squares, you cannot reach any configuration where the number of black special squares is odd.

More Frames on a Smaller Board

This puzzle was brought to my attention by Giulio Genovese, whose Putnam coach Vladimir Chernov found it in the book *Noviye Olimpiady po Matematiker*, Phoenix Press, Rostov-on-the-Don, 2005.

Like the previous chessboard problem, this one involves finding a miraculous "invariant." What follows is a thoroughly inadequate attempt to justify a successful line of reasoning.

Of course, you only need to consider the 2×2, 3×3, and 5×5 boxes since the other two can be made up of smaller ones.

As we have noted earlier, it pays on a problem like this to check whether the number of things you *can* do is adequate to cover the number of things you *have* to do. Here we can regard the grid numbers as being numbers modulo 3 (0, 1, or 2, with $2 + 1 = 0$); hence, there are $3^{6^2} = 3^{36}$ grid numberings to worry about. At each possible position, each type of box can be placed or not placed or placed *twice*; three times would accomplish nothing. There are 5^2 places to put a 2×2 box, 4^2 for the 3×3, and 2^2 for the 5×5, so altogether there are $2^{25} \times 2^{16} \times 2^4 = 2^{45}$ things you can do, which is plenty. Apparently, in fact, many of these things you can do must have the same effect, but it is not yet apparent whether we can get any effect we want.

Mathematically speaking, we have a linear map from the vector space $\mathbb{Z}_3^{2^{45}}$ to the vector space $\mathbb{Z}_3^{3^{36}}$ and we want to know whether the map is "onto." (Being able to get from the all-0 configuration to any arbitrary configuration is equivalent to the reverse.)

If the answer is "yes," then we should be able to get from all 0's to a configuration of all 0's except for one 1 in any chosen location. Moreover, if we can do that, we can solve the problem because we can do this for every location that needs a 1, and twice for every location that needs a 2. So, as in working out (from scratch) how to manipulate a Rubik's Cube™, we look for configurations of boxes that make few changes. An example: Start with two 3×3 boxes in diagonally adjacent positions, so that they overlap in a 2×2 subsquare. Now if you add two 2×2 boxes at each corner of the 4×4 picture, everything cancels except two diagonally opposite corner positions. Thus, we can, if we want to, increment two positions one of which is three steps from the other along a diagonal line.

But it's hard to see how to change a single value in an arbitrary position. Now let's change horses. If it's *not* the case that we can obtain any configuration, there should be an *invariant*: some number associated with a configuration that no application of a box can change. In a linear problem like this one, that invariant should itself be a linear function of the position. This means that there should be two subsets A and B of the positions such that if you add up the values in A and twice the values in B (or, equivalently in our mod-3 arithmetic, if you add the values in A and subtract the values in B) then this would give you the invariant.

The construction above tells us that if some position is in A then the position (there's always just one such) which is three steps away along a diagonal must be in B, and vice versa. Starting from that observation, and knowing that we need in every possible box an equal number (modulo 3) of positions in A and positions in B, we can arrive in a somewhat deliberate manner at the miraculous pattern in Figure 24, in which points in A are labeled with "+" and points in B by "−". Thus the claim is: the sum of the values in the "+" positions, minus the sum of the values in the "−" positions, cannot change. It follows that we cannot get from any position where this value is not 0 to one in which all the values are 0.

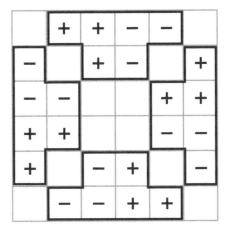

Figure 24. The sets A (squares with a "+" sign) and B ("$-$").

A Simple Bluff

This puzzle was suggested by Jeremy Thorpe and Louise Foucher of Caltech, but similar simple bluffing games can be found in the literature.

We note that Louise cannot lose by raising if she does have a spade, so the "pure" strategies available to her are

- "honest": raise if she has a spade,

- "brazen": always raise.

Jeremy, similarly, has the options when Louise raises of

- "timid": always fold,

- "brave": always call.

The probability of drawing a spade is 1/4; "honest" opposite "timid" yields $1 to Louise 1/4 of the time, $1 to Jeremy the rest, for an expectation of $0.50 to Jeremy. "Honest" opposite "brave" gives $11 to Louise when she has a spade, netting $\frac{1}{4} \times \$11 - \frac{3}{4} \times \$1 = \$2$ to her on average.

On the other hand, "brazen" opposite "timid" wins $1 every time for Louise while "brazen" opposite "brave" costs Louise $\frac{3}{4} \times \$11 - \frac{1}{4} \times \$11 = \$5.50$ in expectation.

If we enter these possibilities into a 2×2 game matrix, we find no dominant strategy for either player. Hence, as intuition would suggest, both players are going to want to use randomized strategies.

From the work of John von Neumann (even before John Nash), we know that there is a *Nash equilibrium* for this game—a pair of strategies such that neither player can improve on his or her strategy given that the other player does not change. Consider what this means from Louise's point of view: if she's not motivated to switch to pure "honest" or pure "brazen," it must be that on average, she doesn't care whether Jeremy calls or folds.

Suppose that Louise decides that when she holds a non-spade, she will bluff with probability p. Against "timid" she expects to win $\frac{1}{4} \times \$1 + p \times \frac{3}{4} \times \$1 - (1-p) \times \frac{3}{4} \times \$1 = \$(\frac{3}{2}p - \frac{1}{2})$. Against "brave" she expects $\frac{1}{4} \times \$11 - p \times \frac{3}{4} \times \$11 - (1-p) \times \frac{3}{4} \times \$1 = \$(2 - \frac{15}{2}p)$.

Louise's indifference means that these quantities are equal, giving us $p = 5/18$; that is, Louise should bluff $\frac{5}{18}$ of the time when she does not have a spade and, of course, always raise when she does. Her expectation regardless of Jeremy's strategy will then be

$$\$ \left(\frac{3}{2} \times \frac{5}{18} - \frac{1}{2} \right) = \$ \left(2 - \frac{15}{2} \times \frac{5}{18} \right) = -\$\frac{1}{12}$$

so that, on average, Louise will lose $\frac{1}{12}$ of a dollar per game.

A little reflection will convince you that if the bet size is increased to more than \$10, Louise will benefit—but never quite to the point where the game is fair; she will always be an underdog. The reason is that she cannot afford to bluff as much as one third of the time when she is spadeless. If she did that, from Jeremy's point of view, the probability that she held a spade given that she raised would be at least one half, thus Jeremy might as well always call; Louise would then break even at best when she raised and lose \$1 when she did not, so she would lose on average. But if $p < \frac{1}{3}$, she'd be losing to Jeremy's "timid" strategy since she'd lose her ante more often than she won Jeremy's.

The fact that the probability of drawing a spade is one quarter is critical here; if it were even a bit better than that (say, if the Queen of Hearts were removed from the deck in advance) then a large enough bet size would turn the tables in favor of Louise.

Going back to the original \$10 bet size, we can also compute Jeremy's equilibrium strategy if we wish (although you weren't asked for that). Suppose that when Louise raises, Jeremy calls with probability q. Then, against her "honest" strategy, he nets $\frac{3}{4} \times \$1 - \frac{1}{4} \times q \times \$11 - \frac{1}{4} \times (1-q) \times \$1 = \$(\frac{1}{2} - \frac{5}{2}q)$. Against "brazen" he reaps $\frac{3}{4} \times q \times \$11 - \frac{1}{4} \times q \times \$11 - (1-q) \times \$1 = \$(\frac{13}{2}q - 1)$. Setting these quantities equal gives $q = \frac{3}{18}$; that is, Jeremy should be calling only $\frac{3}{18}$ of the time. Substituting $q = \frac{3}{18}$ back into the expressions confirms that both give Jeremy an average of $\$\frac{1}{12}$ per game, as must be the case since Louise is losing a like amount.

Chinese Nim

This game is known as Wythoff's Game, as well as Chinese Nim, and was introduced in 1907 paper [60]. It is discussed in several places in Volumes I and II of the classic *Winning Ways for your Mathematical Plays*, by Elwyn R. Berlekamp, John H. Conway, and Richard K. Guy [4]. The connection below to the earlier puzzle Steadfast Blinkers is observed in Serge Tabachnikov's lovely book, *Geometry and Billiards* [56]. However, neither of the two books provides a derivation of the winning strategy.

Each position $\{x, y\}$ in Alex and Beth's game is either winning or losing for the player who plays next, assuming best play by both players. As in classical Nim, it's easiest to try to characterize the losing positions, as there are fewer of those.

Once you know the losing positions, the correct strategy can be derived. If Alex, for instance, is in a winning position, it must be that he can in one move reduce to a losing position for Beth. If he is in a losing position then all he can do is hope for Beth to make a mistake, or to accept his gracious offer to let her play first. So, you can deduce strategy from the list of losing positions, but it seems you are in a circular trap: don't you need to know the right strategy in order to compute the losing positions? Fortunately, since the number of beans is always decreasing, you can start from the bottom and bootstrap your way up.

Any position with one pile empty, or both piles of the same size, is automatically a winning position. It's not hard to deduce that the simplest losing position is $\{1, 2\}$. After that, you can work out that $\{3, 5\}$, $\{4, 7\}$, and $\{6, 10\}$ are losers as well. What's the pattern?

Let $\{x_1, y_1\}, \{x_2, y_2\}, \ldots$ be the losing positions for the first player (not counting $\{0, 0\}$), with $x_i < y_i$ and $x_i < x_j$ for $i < j$. Notice that you cannot have $x_i = x_j$ for $i \neq j$ because then Alex could play to reduce the larger of y_i and y_j to the smaller, leaving Beth with a losing position, a contradiction.

Some thought will lead you to conclude that, given $\{x_1, y_1\}$ up to $\{x_{n-1}, y_{n-1}\}$, x_n is the least positive number not among $\{x_1, \ldots, x_{n-1}\} \cup \{y_1, \ldots, y_{n-1}\}$, and $y_n = x_n + n$. Notice that this forces y_n to be a higher number than any in the set $\{x_1, \ldots, x_{n-1}\} \cup \{y_1, \ldots, y_{n-1}\}$.

The proof is by induction on n. You have already seen that x_n can't be among the numbers in $\{x_1, \ldots, x_{n-1}\} \cup \{y_1, \ldots, y_{n-1}\}$ and also that there can't be more than one y_n to go with this x_n, so all you need to do is show that this $\{x_n, y_n\}$ really is a losing position for Alex.

If $\{x_n, y_n\}$ were a *winning* position for Alex, it must be that he can reduce it to $\{x_i, y_i\}$ for some $i < n$; but he cannot get to this position by reducing the smaller pile or by reducing both piles by the same amount, because that would leave the difference of the two piles at n or more. Nor can he get there by reducing the larger pile, because then he would get another y for the same x. Thus, $\{x_n, y_n\}$ is indeed a loser.

You now have the means to generate as long a list as you like of losing positions. From this Alex's strategy is easy to work out. If he is faced with $\{x_i, y_i\}$, he removes a bean or two and hopes for an error. If he sees $\{x_i, z\}$ for $z > y_i$, he reduces z to y_i. If he sees $\{x_i, z\}$ with $x_i < z < y_i$, thus the difference $d = z - x_i < i$, he takes from both piles to get down to $\{x_d, y_d\}$ (if $z = y_j$ for some $j < i$, he also has the option of just reducing x_i to x_j). If he sees $\{y_i, z\}$ with $y_i \leq z$, he can reduce z all the way down to x_i, and may have other options as well.

But it might take a while to generate enough losing positions to decide what to do with thousands of beans in each pile. Is there a more direct way to characterize the losing positions?

Well, you know that for each n, x_n is somewhere between n and $2n$, because it is preceded by all the x_i's for $i < n$ and *some* of the y_i's. It is reasonable to guess that x_n is approximately equal to rn, for some ratio r between 1 and 2. If so, y_n would be approximately $rn + n = (r+1)n$.

If this holds up, it follows that the n x_i's between 1 and x_n are more or less evenly distributed, and therefore a fraction $r/(r+1)$ of them will have their corresponding y_i below x_n. Thus, there are about $nr/(r+1)$ y_i's below x_n, together with the n x_i's, adding up to x_n numbers in all; making an equation out of this gives

$$n + n\frac{r}{r+1} = nr,$$

which gives us $r + 1 = r^2$, $r = (1 + \sqrt{5})/2$, the familiar "golden ratio."

You might now make the brilliant observation that since r is irrational and $\frac{1}{r} + \frac{1}{r^2} = 1$, r and $r^2 (= r + 1)$ can play the role of p and q in the solution to Steadfast Blinkers from Chapter 3. You saw there that every positive integer can be represented uniquely *either* as $\lfloor pm \rfloor$ for some integer m or $\lfloor qn \rfloor$ for some integer n.

This suggests that perhaps x_n is exactly $\lfloor rn \rfloor$ and y_n exactly $\lfloor r^2 n \rfloor$. Certainly these values have the desired property that each x_n is the least positive number not among $x_1, \ldots, x_{n-1}$ or $y_1, \ldots, y_{n-1}$, since otherwise there's no way to go back and get it. All that remains is to verify that $\lfloor r^2 n \rfloor - \lfloor rn \rfloor = n$, but this is easy: $r^2 n - rn$ is exactly the integer n, so the difference of their floors must also be n. Done!

For fun let's use this to find Alex's move in the example positions. Note that $12,000/r$ is a fraction under 7417, and $7417r = 12,000.9581\ldots$ so 12,000 is an x_i, namely x_{7417}. The corresponding y_{7417} is $\lfloor 7417r \rfloor = 19,417$ so if the other pile has 20,000 beans, Alex can win by taking $20,000 - 19,417 = 583$ beans away from it. If there are only 19,000 beans in the other pile, Alex can win instead by reducing the piles simultaneously to $\{x_{7000}, y_{7000}\} = \{11,326, 18,326\}$. Since 1900 happens to be a y_j, namely y_{2674}, Alex can also win by reducing the x-pile to $x_{2674} = \lfloor 2674r \rfloor = 4326$.

New Visits to Old Friends

Should auld acquaintance be forgot
And never brought to mind?
                    —Robert Burns (1759–1796)

Like fine wine puzzles can improve with age, acquiring new, exciting versions and sometimes better solutions to old versions. In this chapter are some brainteasers that may seem familiar; but even if you think you recall one of these quite well, you may not be prepared for the new twists you see here!

One of the most memorable characters of puzzledom is a logician who likes to vacation in the South Seas. If you believe Martin Gardner (see, e.g., his first *Scientific American* collection [27]), this logician is constantly getting lost and having to ask locals for directions.

Three Natives at the Crossroads

Martin Gardner's logician is again visiting the South Seas and is as usual at a fork, wanting to know which of two roads leads to the village. Present this time are three willing natives, one each from a tribe of invariable truthtellers, a tribe of invariable liars, and a tribe of random answerers. Of course, the logician doesn't know which native is from which tribe. Moreover, he is permitted to ask only two yes-or-no questions, each question being directed to just one native. Can he get the information he needs? How about if he can ask only *one* yes-or-no question?

We proceed next to a famous and elegant geometry puzzle, which, I am embarrassed to say, appeared on page 46 of my previous book [59] with a faulty answer.

Reprise of the Three Circles

The "focus" of two circles is the intersection of two lines, each of which is tangent to both circles, but does not pass between them. Thus, three circles

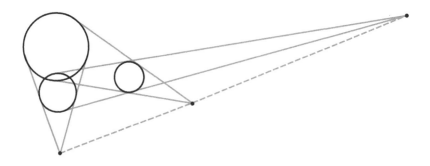

Figure 25. Three circles and their allegedly collinear foci.

of different radii (but none contained in another) determine three foci (see Figure 25). Prove that the three foci lie on a line.

Comment: The solution from [59] entailed the erecting of three spheres, each with one of the given circles as its equator, then considering a plane tangent to the three spheres. However, Jerome Lewis, Professor of Computer Science at the University of South Carolina Upstate, pointed out quite correctly that there may not be such a plane! For example, suppose that two of the given circles are large and between them is a smaller circle.

Remarkably, the proof can be rescued without giving up the nice idea of going to three dimensions. Can you find a way?

The next one is a powerful representative of a class of puzzles based on knowledge about knowledge.

The Dot-Town Suicides

Each resident of Dot-town carries a red or blue dot on his (or her) forehead, but if he ever thinks he knows what color it is he kills himself. Each day the residents gather; one day a stranger comes and tells them something— *anything*—nontrivial about the number of blue dots. Prove that eventually every resident kills himself.

Comment: Nontrivial means that there is some number of blue dots that would make the statement true, and some other number that would make it false. We do *not* assume, however, that what the stranger tells them is true! Alas, the residents of Dot-town are notoriously gullible and will believe anything they hear unless their own eyes tell them otherwise.

Perhaps you remember The Infected Checkerboard, a wonderful puzzle in which your job is to prove that you can't infect an $n \times n$ board starting with fewer than n sick squares; a square becomes infected when two or more of its (orthogonal) neighbors is infected. Showing that n sick squares suffice was the easy part; just take the main diagonal.

What happens when you raise the dimension?

Infected Hypercubes

An infection spreads among the n^d unit hypercubes of a d-dimensional $n \times n \times \cdots \times n$ hypercube, in the following manner: if a unit hypercube has d or more infected neighbors, then it becomes infected itself. (Neighbors are orthogonal only, so each little hypercube has at most $2d$ neighbors.)

Prove that you *can* infect the whole hypercube starting with just n^{d-1} sick unit hypercubes.

Puzzles in which prisoners are fitted with red or blue hats, then get to see their fellow prisoners' hat colors and must guess their own, have caused something of a sensation. In the version that sparked a *New York Times* article [50], each prisoner had an option of whether to guess or pass; and all prisoners were to be executed unless at least one chose to guess, and all who guessed were correct. As usual, the prisoners were permitted to conspire beforehand but could not communicate in any way once the hats were exposed.

In that version, one might think that the prisoners could do no better than plan to have one person guess and everyone else pass, accepting a 50% chance of survival. Yet, in fact they can do much better; n prisoners can cut their probability of execution down to about $1/n$.

For those of you who find such problems off-puttingly fanciful, brace yourselves: you're about to see some variations that will make the old versions seem realistic by comparison.

Hats and Infinity

Each of an infinite collection of prisoners, numbered $1, 2, \ldots,$ is to be fitted with a red or blue hat. At a prearranged signal, all the prisoners are revealed to one another, so that everyone gets to see all his fellow prisoners' hat colors—but no communication is permitted. Each prisoner is then taken aside and asked to guess the color of his own hat.

All the prisoners will be executed unless *only finitely many* guess wrongly. The prisoners have a chance to conspire beforehand; is there a strategy that will ensure survival?

Comment: Note that this is a simpler problem than the one described at the beginning of the chapter: there is no option to pass, nor any parameter, nor any issue of probability; we are asking for a perfect solution. We do implicitly assume, however, that each prisoner can actually apprehend the entire sequence of hat colors that he sees and somehow process all that information in finite time to arrive at his guess.

All Right or All Wrong

This time the circumstances are the same but the objective is different: the guesses must either be *all right* or *all wrong*. Is there a winning strategy?

Comment: The finite version of this one is pretty easy: the prisoners might, for example, decide in advance that each will guess his own hat color under the assumption that the total number of red hats is even. Then, if it *is* even, they'll all be right; otherwise, they'll all be wrong. But in the infinite case, if the number of red hats is infinite, how can they decide if there are an even number of them?

We return now to finitely many prisoners, but with hats replaced by numbers.

Numbers on Foreheads

This time, each of 10 prisoners will have a digit between 0 and 9 painted on his forehead (they could be all 2's, for example). At the appointed time each will be exposed to all the others, then taken aside and asked to guess his own digit.

In order to avoid mass execution, at least one prisoner must guess correctly. As usual the prisoners have an opportunity to conspire beforehand; show that there is a scheme by means of which they can ensure success.

The Color-Blind Prisoner

Unfortunately, the prisoners of the previous puzzle discover that that there is a problem: one of them (named Shrek) has green skin. The digits will be painted in red, and prisoner Mike is red-green color blind. Thus, Mike will have to make his guess on the basis of only 8 visible digits. The rest of the prisoners, including Shrek, will still be able to see all 9 digits other than their own. *Your* task is to prove that the prisoners can no longer avoid some possibility of execution.

Finally, we combine numbers *and* hats in our last prisoner puzzle.

Numbers and Hats

Each of n prisoners has a distinct real number written on his forehead, so that each can see the others' numbers but not his own. As usual no communication is permitted after the viewing, but then each prisoner must independently choose a hat (either red or blue) for himself.

The goal is for the hat colors to alternate in the order determined by the real numbers.

How can the players, who are permitted to conspire beforehand, maximize the probability of success?

We end our visits to old puzzle friends with one that dates back at least to the middle of the nineteenth century but was not solved until 2006. We won't ask you to prove your solution is correct (though you're of course welcome to try)—it took five professional mathematicians to wrap this one up, and even now the answer is known only up to a constant factor. But, guessing a construction is a great test of intuition.

Tower of Bricks

How far can a stack of n bricks be made to lean over the edge of a table?

Comment: You may assume that the bricks are frictionless, homogeneous rectangular solids of length 1, and stacked horizontally in a common vertical plane. But *don't* assume that there is only one brick at each level.

Sources and Solutions

Three Natives at the Crossroads

This version of "logician at the crossroads" came to me from two mathematical physicists, Vladas Sidoravicius and Senya Shlosman. It seems impossible to deal with the random native, but you can.

The idea is, you need to be sure the *second* native you query is not the random answerer. This is necessary because you're not going to know the right road after the first question, and if the second responder is random you might not learn any more.

On the other hand, this objective is sufficient, because you can then use the traditional one-native query as your second question: namely, something like "If I were to ask you whether Road 1 goes to the village, would you say yes?"

To attain the objective, you'll need to ask Native A something about Native B or Native C, then use the answer to choose between B and C. Here's one that works: "Is B more likely than C to tell the truth?"

Curiously, if A says "yes" you pick C, and if he says "no" you pick B! If A is the truth-teller you want next to query the companion who is *less* likely to tell the truth, namely, the liar. If A is the liar you want the *more* truthful of his companions, namely, the truth-teller.

Of course, if A is the random answerer it doesn't matter which of B and C you turn to next.

In Martin Gardner's columns it was pointed out that in the original one-native problem, the logician can get to the village even if he has forgotten which of the native words ("pish" and "tush," allegedly) means "yes" and which means "no." Readers seeking further challenge can attempt to modify the above protocol similarly.

If the random answerer operates simply by flipping a mental coin to decide whether to say "yes" or "no," it is of course impossible to determine which road leads to the village with only *one* question. But, suppose that he instead decides between telling the truth and lying, and then backs his decision with careful analysis. Anupam Jain of the University of Southern California suggests that the logician ask:

> *Out of the other two guys, if I pick the one whose response's truthfulness will least likely match your response's truthfulness and ask him if Road 1 goes to the village, will he answer "Yes"?*

The claim is that if the answer is No, then Road 1 is the right road; otherwise it's Road 2.

Suppose that Road 1 is the right road.

The critical case is when the logician asks this question to the random answerer. If the random answerer has decided to tell a lie for this question, then the truthful guy's response's truthfulness will be least likely to match his response's truthfulness. The truthful guy will say "Yes," and since the random answerer has decided to lie, he will say the opposite and thus answer "No."

If the random answerer has decided to tell the truth for this question, then the *liar's* response's truthfulness will be least likely to match his response's truthfulness. The liar would say "No" and since the random answerer has decided to tell the truth, he will also say "No."

If the logician is addressing the truth-teller, it will be the liar whose truthfulness will be least likely to match his truthfulness, and he will say what the liar would have said: "No."

Similarly, if Road 2 is the right road, all answers will be "Yes."

Reprise of the Three Circles

The three circles of this puzzle are sometimes called "Monge circles."

The following proof, attributed by cut-the-knot.org to Nathan Bowler of Trinity College, Cambridge, works by erecting *cones* instead of spheres on top of the circles. Call them C_1, C_2, and C_3, and let them all be "right" cones—that is, they support 90° angles at their apices. (Actually, we only need them all to have the same angle.) Each pair of cones determines two (outside) tangent planes, say P_1 and Q_1 (for cones C_2 and C_3), P_2 and Q_2 (for cones C_1 and C_3), and finally P_3 and Q_3 (for cones C_1 and C_2).

Each pair of planes P_i, Q_i intersect in a line L_i that passes through the apex of both tangent cones, as well as through the point where the corresponding circle tangents meet. Thus, in particular, L_1 and L_2 both meet at the apex of C_3, L_1 and L_3 at the apex of C_2, and L_2 and L_3 at the apex of C_1. Hence, the three lines of intersection are coplanar (all lie on the plane determined by the three apices); the intersection of that plane with the original plane of the circles is a line through the three foci, and we are done!

The Dot-Town Suicides

This unusually general knowledge-about-knowledge puzzle came to me from Nick Reingold, of AT&T Labs. Various more specific versions (often in poorer taste even than this one) have existed for many decades.

Many readers will already have seen special cases of this puzzle amounting to the case when all residents have blue dots and the stranger merely says "There is at least one blue dot."

The big surprise here is not just that *anything* the stranger says is disastrous; it's that *even when he says something that everyone knows is false* the residents of Dot-town are doomed by the statement. We will prove this below, but it may be more persuasive to look first at a small special case and see how this works.

Suppose that there are just three residents, all with blue dots, and the stranger tells them "all dots are red." Everyone sees that he is lying, but Resident 1 thinks: Suppose that my dot is red; then Resident 2 sees my red dot and wonders whether Resident 3 is seeing two red dots. If so, thinks Resident 2, Resident 3 will believe the stranger and kill himself tonight, even though his dot is blue. If he doesn't do that, Resident 2 will conclude correctly that Resident 3 saw only one red dot; and will kill himself on night 2. Since neither of these events transpires, Resident 1 concludes that Resident 2 did not see a red dot, thus Resident 1 knows his dot is blue and commits suicide on the third night.

To do a proof of the general case, we need some notation. Let $S \subset \{0, 1, \ldots, n\}$ be the set of numbers x with the property that if there are x blue dots among the n residents of Dot-town, then the stranger's statement is true; our nontriviality assumption tells us that S is a proper, nonempty subset. Let b be the actual number of blue dots, which may or may not be in S.

For Resident i, let B_i be the set consisting of the possible numbers of blue dots, from i's point of view. Prior to the stranger's visit, $B_i = \{b_i, b_i + 1\}$ where b_i is the number of blue dots i sees among his compatriots.

If at any point B_i drops to one value, Resident i is cooked. This will happen immediately if $|B_i \cap S| = 1$, but it will also happen the night after any suicides occur. To see this, we observe first that all residents with the same dot color will behave identically, since they all see the same number of dots. Thus, if Resident i sees that anyone has committed suicide, he deduces (correctly) that that person's dot color is different from his own; he therefore knows his own color and is doomed.

Given S and b, let $d(b)$ be the number of steps (increments or decrements by 1) needed to get from b across the border of S; in other words, $d(b)$ is the least k such that $b + k$ or $b - k$ is in $\{0, 1, \ldots, n\}$ but in S (if b is not) or out of S (if b is in S).

For example, if $n = 10$ and $S = \{0, 1, 2, 9, 10\}$, then $d(0) = 3$, $d(1) = 2$, $d(2) = d(3) = 1$, $d(4) = 2$, $d(5) = d(6) = 3$, $d(7) = 2$, and $d(8) = d(9) = d(10) = 1$.

We have already noted, in effect, that if $d(b) = 1$ then there will be suicides already on the first night. We now claim that, more generally, the first suicides will occur exactly on night $d(b)$.

The proof is by induction on $d(b)$. Suppose that it is true whenever $d(b) < t$, and now let $d(b) = t > 1$. The day after night $t - 1$, since no suicides have yet occurred, everyone will learn that $d(b) \geq t$. However, if $d(b) = t$ then either $d(b - 1)$ or $d(b + 1)$ must be equal to $t - 1$. If the former, then those residents with blue dots—who see that the number of blue dots is either b (the actual number) or $b - 1$—can rule out $b - 1$ and are toast. If the former, it is the red-dotted folks who can rule out $b + 1$ and must do themselves in. Finally, if $d(b - 1) = d(b + 1) = t - 1$, then nobody survives the night.

Since $d(b)$ is at most n, the proof tells us that everyone will have perished by the nth night. We can also see that they last that long only in four extremal cases: when $b = 0$ and $S = \{n\}$ or $\{0, 1, \ldots, n-1\}$, and when $b = n$ and $S = \{0\}$ or $\{1, 2, \ldots, n\}$. One way to say this is that survival time is maximized when the stranger either makes the least informative correct statement or tells the most outrageous lie.

It is perhaps worth noting also that the definition of $d(b)$ makes no distinction between S and its complement; from this it follows that it makes no difference whether the stranger says "X" or "Not X," the residents of Dot-town will behave exactly the same way in either case.

You might reasonably wonder whether the Dot-town residents, knowing that a stranger is coming and might break the manifestly justifiable no-talk-about-dot-colors taboo, can organize some defense. For example, everyone who knows the stranger is lying jumps up and says so. Alas, a little thought will show you that neither this nor any similar strategy can save the town.

A fragile lot, these Dot-towners. Oddly, though, their very fragility could save them; Steve Babbage, a manager and cryptographer with Vodafone, points out that if they begin to worry that a suicide was not caused by knowing one's dot color—but perhaps because some Dot-towner "has finally cracked under the strain of living in such a ludicrous environment"—then under certain circumstances the rest of the town may yet survive the stranger's incursion.

Infected Hypercubes

The proof that initially infecting at least n^{d-1} sick unit hypercubes ("sites" for short) is *necessary* is a straightforward generalization of the two-dimensional case, in which we observed that the perimeter of the infected area cannot increase. Here, we replace perimeter by the $(d-1)$-dimensional area of the infected region's surface. When a new site becomes infected, at most d of its $(d-1)$-dimensional faces are added to the surface of the infected region, while at least d faces are removed (those that separate the site from its already infected neighbors). So, again, this area cannot increase. Its final value is the surface area of the big hypercube, which is $2d \times n^{d-1}$. If there are initially k infected sites, the initial surface area cannot be greater than $k \times 2d$ since each site has $2d$ unit-area faces. It follows that k must be at least n^{d-1}.

But, this time, it's not so obvious how to choose the initial n^{d-1} sites to infect. Matt Cook and Erik Winfree of Caltech found a way that seemed to work, but they couldn't prove it; finally their colleague Len Schulman came up with the remarkable proof below (sent to me by Winfree).

First, here's Matt and Erik's construction. Label the sites by vectors $(x_1, x_2, \ldots, x_d)$, with $x_i \in \{1, 2, \ldots, n\}$, so that two sites will be neighbors if all their coordinates are the same except in one position, where their values differ by 1.

Choose any integer k, and infect all sites such that $\sum_i x_i \equiv k \mod N$. These sites form a "diagonal subspace" that is cut up into many pieces. It fills up in a really weird way, quite differently depending on the value of k. Usually, it seems to barely make it, relying on many apparent coincidences that allow the growth to continue. Very different from the two-dimensional case! It seems remarkable that the process manages to infect the whole big hypercube.

To prove it really works, Schulman considered the following game, which represents the forces preventing infection as a demon adversary trying to trap you, the infector. Let us choose a k and begin the infection as above.

In the game, the demon begins by putting you at $x = (x_1, \ldots, x_d)$. Now repeat: The demon chooses dimension index i, and you move either forward or backward in that dimension (if the site's ith coordinate is 1 or n, you won't have a choice). You win if you can reach some x such that $\sum_i x_i \equiv k \mod N$; the demon wins if she can keep you wandering forever.

We now claim that if you have a strategy that guarantees a win, then the d-dimensional hypercube will indeed be fully infected.

To prove this, we first refine the claim to state that if you can win starting from x, then x itself will become infected. Note that, from x, the demon might choose any direction i. A winning strategy must work for all d such possibilities. This implies that your strategy also wins if started at any of the d neighbors of x to which you were preparing to move. By induction (over the number of steps to a win), all these d neighbors of x can be infected, thus x can as well. The base case of the induction is when the starting point x has coordinates summing to k modulo n, in which case it is of course already sick.

Now all we need is to provide you with a winning strategy; Schulman calls what follows the "wheelbarrow algorithm." For any site x, let x^* be the number $k + \frac{1}{2} + \sum_i x_i \mod N$. After the demon chooses some coordinate i, if $x_i > x^*$, you must decrement x_i (thus x^* goes down as well, though possibly around the corner from $\frac{1}{2}$ to $n - \frac{1}{2}$). If, on the other hand, $x_i > x^*$, you increment x_i so x^* goes up too, perhaps from $n - \frac{1}{2}$ to $\frac{1}{2}$. But wait a minute—if you ever get to a site where $x^* = \frac{1}{2}$, you've won the game!

It follows that the move prescribed by the algorithm is always legal: You will never be asked to move to a site where $x_i = 0$ or $x_i = n + 1$ unless you have already won.

We now assert that the demon can't force you to cycle. Suppose to the contrary that x cycles forever, and let I be the set of indices chosen infinitely often by the demon. We may as well assume that you are past the point where any index not in I will ever be chosen. Let y be the biggest value of x_j ever encountered for any $j \in I$. Let J be the set of indices in I that are at the moment exhibiting that maximum value y.

If it ever happens that $x^* > y$, then you will be incrementing at every step, pushing x^* up until it snaps around the corner to $\frac{1}{2}$ and you win. Therefore, it must be that x^* is always below y. But then, whenever the demon chooses a $j \in J$, x_j must decrease to $y - 1$. It follows that eventually J will disappear, leaving you forever with a smaller maximum value y. This can't go on forever, so we have our contradiction.

It follows that the wheelbarrow algorithm will win the game for you, no matter where the demon starts you off or how she chases you around. The existence of a winning strategy means that the infection really does capture the whole big hypercube, and we are done.

Hats and Infinity

The answers "Yes, there is a winning strategy" and "No, there is not" are both correct! How can this be?

To the best of my knowledge, this lovely hat puzzle was devised jointly by Yuval Gabay and Michael O'Connor (then graduate students at Cornell University), but the solution was already implicit in the work of Fred Galvin of the University of Kansas. Christopher Hardin (Smith College) and Alan D. Taylor (Union College) then included it in a paper for the *American Mathematical Monthly* [36]. Stan Wagon wrote it up as a Macalester College Problem of the Week; additional nice observations about this and the next version were made by Harvey Friedman (Ohio State), Hendrik Lenstra (Universiteit Leiden), and Joe Buhler (Reed College). It was the last of these, and (independently) Matt Baker of Georgia Tech, who communicated the puzzle to me. And all this is a considerable simplification of the puzzle's history—please forgive me if your name was left out!

Let us consider for a moment what might happen if only finitely many of the prisoners are given red hats. Then, all the prisoners will see this, and if they so agreed in advance, will all guess "red"—and of course only finitely many will be wrong.

The same scheme can also be applied if only finitely many of the hats are blue—or, for example, if only finitely many of the odd-numbered hats are red and finitely many of the even-numbered hats are blue. In fact, let's go even further: if the hat sequence is eventually *periodic*, then everyone agrees to guess as if the sequence were periodic from the start.

To put it another way, we can translate the hat sequence into the binary representation of a real number r in the unit interval $[0,1]$, regarding (say) blue as 1, red as 0. "Eventually periodic" means that after some point there is a finite 0-1 pattern which repeats forever; this is equivalent to saying that r is rational. For example, the sequence $100101001010101010\ldots$, where 01 repeats forever, is such a number; it happens to be the fraction 7/12. If it occurred then every odd-numbered prisoner would guess "blue" and every even-numbered prisoner "red", and all except numbers 3, 4, 5, 6, and 7 would be correct.

So, the prisoners will be fine if the hat sequence is rational, but why limit the strategy to rationals? Maybe the sequence will differ in only finitely many places from the binary representation of π, in which case the prisoners can agree to guess as if the hats represented π exactly.

What the prisoners really need is to divide up all the possible hat sequences into "classes" having the property that within each class, two sequences differ in only finitely many places. Then, the prisoners agree in advance on a *representative* from each class, that is, some particular member of the class. If a sequence from that class is observed, then everyone guesses as if the actual sequence were equal to the agreed-upon representative.

Mathematically, what we are doing is defining two sequences to be (say) *neighbors* if they differ in only finitely many places. We observe that (1) any r is a neighbor of itself; (2) if r is a neighbor of s then s is a neighbor of r; and (3) if r is a neighbor of s and s is a neighbor of t, then r is a neighbor of t. That means our notion of neighbors is what mathematicians call an *equivalence relation*. This in turn means that there really is a partition of the set of all hat sequences into classes, so that within each class any two sequences are neighbors, but any two sequences in different classes are *not* neighbors.

So far so good, but now we hit a sticky point. Most of these classes will not come equipped with some natural representative (like π); the prisoners will have to make many arbitrary choices. The mathematical axiom that says they can do this is called the *Axiom of Choice*. If the Axiom of Choice holds, then the prisoners can apply it to decide on a representative of each class. Then, when the hats are observed, the prisoners will all agree on which class the hat sequence belongs to (if they wish, they only need check all the hats belong to prisoners with higher numbers than their own). They then all guess that their own hat color conforms to that predicted by the agreed-upon class representative, and automatically only finitely many of the prisoners will be wrong. Voila!

But, *is* the Axiom of Choice true? Most practicing mathematicians routinely assume that it is, but the fact is, if the basic rules of mathematics are consistent *with* the Axiom of Choice, then they are also consistent *without* the Axiom of Choice. So it is just as valid to assume that the Axiom of Choice fails as it is to assume that it holds.

And if the Axiom of Choice fails, the prisoners may be stuck. The aforementioned Hardin and Taylor, and separately Harvey Friedman, have shown that it is consistent with the standard axioms of mathematics that *there is no solution* for the prisoners. Even worse, any proposed solution *will* fail unless, in a certain sense that can be made mathematically precise, the prisoners are outrageously lucky. So, if you are likely to become a prisoner, you'd better pack the Axiom of Choice in your bag.

Do *you* believe in the Axiom of Choice? Consider: the set of possible winning strategies for our proposed solution is the product of all the equivalence classes discussed above; if there's no solution, that means that the product of an infinite number of nonempty (indeed infinite) sets is *empty*. On the other hand, the famous Banach-Tarski paradox assures us that we can use the Axiom of Choice to cut up a ball into five pieces and reassemble them to make two balls, each identical to the original!

My own summary of the situation is this: while neither the Axiom of Choice nor its negation can be disproved, either can be made to look ridiculous.

All Right or All Wrong

If you followed the (Axiom of Choice–based) solution to the previous puzzle, you're in good shape here. The prisoners agree as before on a representative from each equivalence class, and here they *also* agree in addition that they will guess their own hat colors under the assumption that the number of positions in which the hat sequence disagrees with the chosen representative is even. As in the finite case, if that number *is* even, all the prisoners will be right; otherwise, all wrong.

Interestingly, mathematicians who teach or do research in algebra typically solve this problem the following way, instead. Identify each hat color with the set $\{0, 1\}$ as before, which we may think of as the two-element group $\mathbb{Z}_2$. The *sum* Σ of infinitely many copies of $\mathbb{Z}_2$ is the set of all sequences of 0's and 1's only finitely of which are 1's (blue hats); there is a natural homomorphism (mapping) from Σ to $\mathbb{Z}_2$ which gives 0 if the number of 1's is even and 1 if odd. A standard "extension theorem" of algebra allows us to extend this homomorphism to the *product* of the $\mathbb{Z}_2$'s, which is the set of *all* $\{0, 1\}$-sequences. The prisoners then agree to guess assuming that the value of this homomorphism is (say) 0.

Of course, *proving* the extension theorem requires the Axiom of Choice, so this proof is really no different from the other.

Does this mean that the Axiom of Choice is also required for the all-right-or-all-wrong version of the infinite hat problem? Well, I thought so, but Teena Carroll (a mathematics graduate student at Georgia Tech) pointed out to me that there's a *much* easier solution which doesn't require the Axiom of Choice or any mathematics at all.

Everyone guesses "green"!!

Numbers on Foreheads

This puzzle came independently from several sources, including Noga Alon of Tel Aviv University. As Noga has himself proved many times, it is useful in many problems to introduce probability even though none is present in the statement. Here, if we assume that the numbers painted on foreheads are chosen independently and uniformly at random, we see that no matter what he does, each prisoner has probability exactly $\frac{1}{n}$ of guessing correctly.

Let the prisoners be numbered from 0 to $n-1$. Since we want the probability that *some* prisoner guesses correctly to be 1, we need the n events "Prisoner k guesses correctly" to be mutually exclusive: in other words, no two can occur. Otherwise, the probability of at least one success would be strictly less than $\frac{1}{n} + \frac{1}{n} + \cdots + \frac{1}{n} = 1$.

To do this it would behoove us to separate the set of possible configurations into n equally likely scenarios, then have each prisoner base his guess on a different scenario. This reasoning may already have led you to the easiest solution: Let s be the sum of the numbers on the prisoners' foreheads, modulo n. Now let Prisoner k guess that $s = k$, in other words, guess that his own number is k minus the sums of the numbers he sees, modulo n.

This will ensure that Prisoner s, whoever that may be, will be correct (and all others wrong).

The Color-Blind Prisoner

Owing to the unfortunate situation with Mike and Shrek, the prisoners are unable to ensure a successful implementation of the scheme above. But that doesn't in itself mean that some other scheme can't come to the rescue.

However, we have seen that a successful scheme must prevent any two prisoners from ever guessing correctly, and that perhaps suggests a problem for Shrek. Suppose that there *is* a successful scheme; we may as well assume that Shrek knows what Mike is supposed to do. Then, since he can see every number that Mike can see, he in fact knows, when the numbers are exposed, exactly what Mike *will* do. And since he can see Mike's number, he might (with probability $\frac{1}{n}$) see that Mike is about to guess correctly.

In that case Shrek must guess incorrectly, but since he doesn't know his own number he has no way of doing so. ("Firing into the air" by, say, guessing $n + 1$, won't work because then the prisoners' success probabilities have no chance to add up to 1.) This contradiction ensures that the prisoners will have to accept some positive probability of failure.

Numbers and Hats

This puzzle was sent to me by Nicole Immorlica, a postdoc at Microsoft Research. Its statement (and solution) is found in a six-author paper [1], where it is used in an explicit construction for high-revenue deterministic auctions.

In fact, the prisoners can guarantee a win against any distribution of the numbers. Before their foreheads are painted, they assign a fixed order to themselves, that is, they number themselves from 1 to n (perhaps alphabetically). After the viewing, for each i, prisoner i assigns new numbers 1 through $i - 1$ and $i + 1$ through n to his fellow prisoners, in the order he (or she) sees given by the real numbers on their foreheads. He then computes how many transpositions of old numbers it would take to get them to where the new numbers are.

Suppose that this ith prisoner later had his own real number revealed to him, and it turned out to be the jth in the full order of all n reals. He would then have $|i - j|$ more transpositions to make in order to complete the permutation σ from all n old numbers to the n new numbers, because i and j have to be swapped and the numbers between them shifted up or down by one.

For example, suppose that $n = 4$ and Prisoner 3 sees 2π, π, and 4π on the foreheads of Prisoners 1, 2, and 4, respectively. Then, he assigns them new numbers 2, 1, and 4. To get these numbers from the old 1, 2, and 4 he just has to switch 2 and 1, a single transposition. If he puts himself in as a 3, he has created the permutation $1234 \rightarrow 2134$. But maybe his own real number is $\pi/2$, so he should have been first, not third, in the order of real numbers. To get the correct permutation $\sigma = 1234 \rightarrow 3214$ he still would need to swap the 3 and the 1, then the 2 and the 3: two more transpositions. Of course, this second stage is hypothetical only, its purpose being to elucidate the argument below.

If σ is an *even* permutation, that is, one which can be realized by an even number of transpositions, then the original permutation done by prisoner i on the remaining numbers is even if $|i - j|$ is even and it is odd otherwise. Of course, if σ is odd (as in the example) the reverse is true.

So, what Prisoner i needs to do is to choose "red" if he counts an even number of transpositions (in his permutation of the $n - 1$ other numbers) and his own old number i is even or if he counts an odd number of transpositions and i is odd. Otherwise, he chooses "blue." In the example i is odd and Prisoner 3 did an odd number (one) of transpositions, so he will choose a red hat.

The effect of this choice if σ happens to be even is that prisoner i will go with "red" just when i and $|i - j|$ are both even or when i and $|i - j|$ are both odd—in other words, when j is even. Thus, in the new order every even-numbered prisoner will be wearing a red hat and every odd one blue.

If σ is odd (as in the example) it will be the odd-numbered prisoners in the new order who are wearing red; either way they have won the game.

Tower of Bricks

The problem of stacking n bricks so as to lean out as far as possible over the edge of a table was posed explicitly in 1923 in the *American Mathematical Monthly* by J. G. Coffin [12], but in other forms as far back as 1850. Martin Gardner helped spread its fame, and it is widely used around the world to introduce students to the harmonic series.

Ironically, however, the famous "harmonic stack" (see Figure 26) is not, despite wide belief, even close to the optimal solution—unless, as is sometimes the case, the puzzle is presented with a one-brick-per-level restriction. Moreover, it is often noted that one needs to know in advance how much overhang is sought; this too is incorrect.

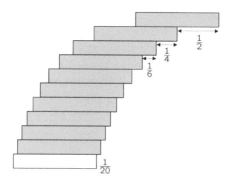

Figure 26. A ten-brick harmonic stack.

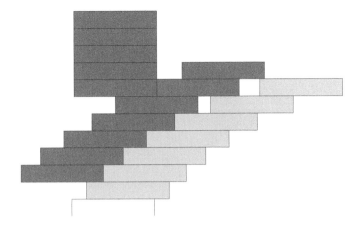

Figure 27. Best possible overhang for 19 bricks.

The harmonic stack is obtained by observing that the top brick cannot extend more than $\frac{1}{2}$ (times the unit brick length) beyond the brick beneath it. If it does so, the center of gravity of the two top bricks is $\frac{1}{4}$ of a brick to the left of the lower brick's extent, so the next-to-top brick cannot extend more than $\frac{1}{4}$ over the brick supporting *it*. Continuing in this manner, we have the kth brick from the top extending by $\frac{1}{2k}$ over the $(k+1)$st, for a total overhang of $\frac{1}{2} + \frac{1}{4} + \frac{1}{6} + \cdots + \frac{1}{2n} = \frac{1}{2}(1 + \frac{1}{2} + \frac{1}{3} + \cdots + \frac{1}{n}) = H_n/2$, where H_n is the nth partial sum of the harmonic series, asymptotically equal to the natural log of n.

Since the harmonic series diverges, you can conclude (correctly) that if you have enough bricks, you can make the overhang as large as you like. But if you do it this way, you do have to know in advance how far out you have to go; otherwise, the placement of the very first brick will limit you to bounded overhang.

It has often been observed, however, that you might do better by using some of the n bricks to counterbalance the others. As recently as December 2005, in a cover article in the *American Journal of Physics* [35], J. F. Hall observed that you can get about twice the overhang (thus, about $\ln n$) with a stack of bricks, each extending out past the one below, counterbalanced by more bricks behind them. Indeed, for up to 19 bricks, such configurations give optimal overhang; see Figure 27 for the $n = 19$ case. However, Hall incorrectly concluded that such configurations (called *spinal* because the bricks supporting the maximally-overhanging brick contain only one per level) are optimal in general.

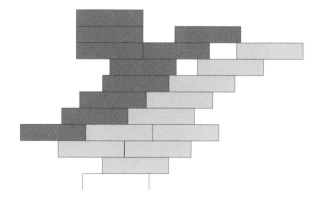

Figure 28. Best possible overhang for 20 bricks.

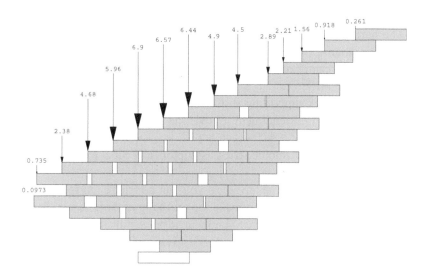

Figure 29. Best possible overhang for 100 bricks.

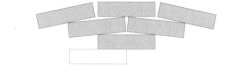

Figure 30. The inverted triangle of three or more layers falls apart.

In fact, in their breakthrough paper [47] in the *Proceedings of the SIAM Symposium on Discrete Algorithms* (January 2006—thus written before Hall's paper appeared), Mike Paterson and Uri Zwick proved that Hall was correct about the overhang achievable with spinal stacks. But, they also showed that spinal stacks cease to be optimal for 20 or more bricks. Even more startling, they presented a different construction that achieves *exponentially better* overhang than previously thought possible.

The actual optimal construction for 20 bricks is pictured in Figure 28; this beats Hall's construction for 20 bricks by only a tiny fraction. However, as you can see from Figure 29, the best configurations begin to look quite far from spinal as n grows. The arrows at the top of Figure 29 represent additional weight supplied by some un-pictured bricks (among the allowed 100) whose exact positions are not uniquely determined.

In fact, for very large n it appears that the best overhang is obtained by configurations that look as if they were cut out of a standard brick wall, in which each brick is centered over the boundary of two touching bricks in the layer beneath. But, the most obvious shapes turn out not to be stable. Inverted triangles (one brick on the bottom, then two, then three etc.) are claimed to be stable in the book *Mad About Physics* by Jargodzky and Potter [38]—highly recommended despite this error—but are in fact unstable as soon as you get to three layers (see Figure 30 to see how they fall apart).

Diamonds (with layers of one brick, up to something, and back to one) are stable up to seven layers but Figure 31 shows what happens after that.

Instead, Paterson and Zwick constructed brick walls that are roughly parabolic in shape, as shown in Figure 32. These are built (and their stability proved) recursively, by piling what they call k-slabs for successively larger k.

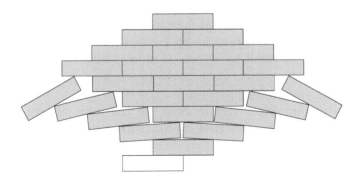

Figure 31. The nine-layer diamond is, alas, unstable.

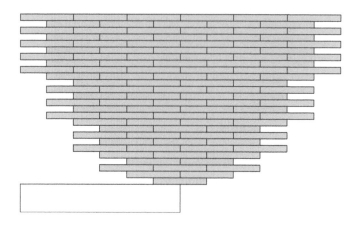

Figure 32. A parabolic brick wall.

A k-slab consists of $2k + 1$ alternating layers of $k + 1$ and k bricks each. The overhang achieved by a parabolic brick wall of n bricks is of order $\sqrt[3]{n}$ (the cube root).[9]

But is this best possible? Diamonds or inverted triangles, if they were stable, would give asymptotically better overhang, of order $\sqrt{n}$ (the *square* root). However, very recently Paterson and Zwick, together with Yuval Peres, Mikkel Thorup, and this writer [48], were able to show that no construction can do better than order $n^{\frac{1}{3}}$.

This doesn't mean that parabolic brick walls are exactly optimal, however; they achieve overhang roughly $\sqrt{3/16}n^{1/3}$, but other constructions may achieve $cn^{1/3}$ for some larger c. Paterson and Zwick believe that the best shape for large n is the "oil lamp" shape of Figure 33.

The parabolic brick wall cannot be built brick by brick—like all the stable configurations pictured above, it is right at the edge of stability and doesn't admit a brick-by-brick construction—but by altering the parabola slightly, one can build it from the table up. Figure 34 shows a modification, still worth overhang of order $n^{1/3}$, that can be made by piling bricks in the indicated order.

Of course, one should keep in mind that real bricks are not perfectly formed and are far from frictionless. So don't try this at home.

---

[9]To say that one function $f(n)$—here, the overhang achieved by n bricks—"is of order $g(n)$" means that there are positive constants c and c' such that $cg(n) < f(n) < c'g(n)$.

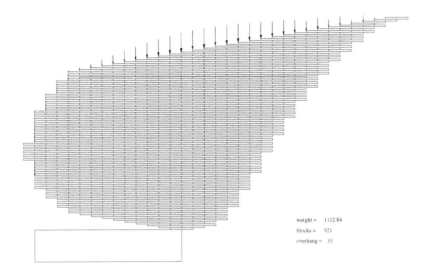

weight = 1112.84
blocks = 921
overhang = 10

Figure 33. An "oil lamp"-shaped brick wall, believed to be nearly optimal.

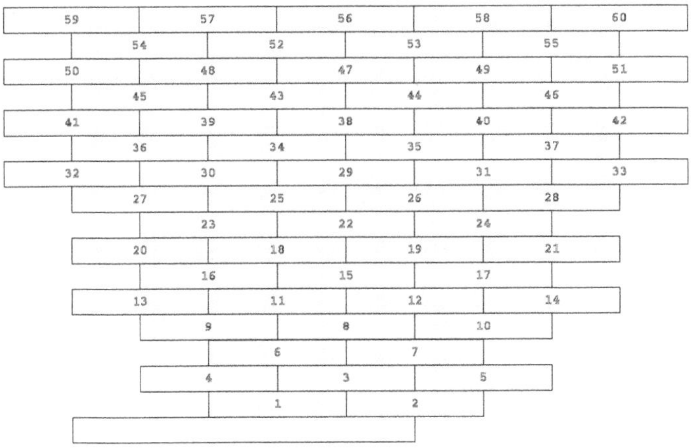

59		57		56		58		60
	54		52		53		55	
50		48		47		49		51
	45		43		44		46	
41		39		38		40		42
	36		34		35		37	
32		30		29		31		33
	27		25		26		28	
	23		22		24			
	20		18		19		21	
	16		15		17			
	13		11		12		14	
	9		8		10			
		6		7				
	4		3		5			
		1		2				

Figure 34. A way to build out to arbitrary overhang.

Severe Challenges

There comes a time when the mind takes a higher plane
of knowledge but can never prove how it got there.

—Albert Einstein (1879–1955)

As if the previous puzzles weren't hard enough, here are a few more toughies. Actually, you might find some of these easier than foregoing puzzles; often what stumps one person is a breeze for another.

Ice Cream Cake

On the table before you is a cylindrical ice-cream cake with chocolate icing on top. From it you cut successive wedges of angle x, where x is arbitrary. Each time a wedge is cut, it is turned upside-down and reinserted into the cake. (See Figure 35.)

Prove that after a finite number of such operations, all the icing is back on top of the cake!

Comment: This problem belongs to the category I call "puzzles you must not have heard correctly." Yes, the angle x can be irrational, in which case you will never cut out the same piece twice. You might in fact have to do cut quite a few wedges—luckily, ice cream cakes are self-healing—but not an infinite number.

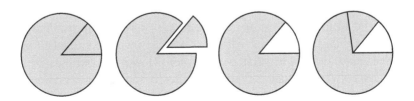

Figure 35. Cutting, inverting, replacing, and cutting again.

Hopping and Skipping

A frog hops down a long line of lily pads; at each pad, he flips a coin to decide whether to hop two pads forward or one pad back. What fraction of the pads does he hit?

Curve and Three Shadows

Is there a simple closed curve in 3-space, all three of whose projections onto axis planes are trees?

Comment: This means that the shadows of the curve, from the three coordinate directions, may not contain any loops. In Figure 36 is a curve that doesn't quite work: two of its shadows are trees but the third contains (indeed, is) a closed curve.

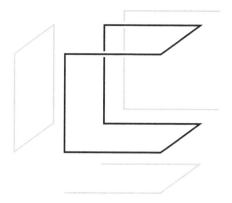

Figure 36. A closed curve with two tree-shadows.

Players and Winners

Tristan and Isolde expect to be in a situation of severely limited communication, at which time Tristan will know which two of 16 basketball teams played a game, and Isolde will know who won. How many bits must be communicated between Tristan and Isolde in order for the former to find out who won?

Comment: This is a problem in *communication complexity*. If Isolde knew who played as well as who won, she could send one bit to tell Tristan whether or not (say) the alphabetically first team won the game. Without that information, she could simply send four bits to identify the winning team, but can they do better?

The next problem is again in communication complexity, but tougher.

Charlie and the Cheaters

Alice and Bob each know all the answers to an n-question true-false exam that Charlie is about to take. Charlie only needs the answer to question k, but neither Alice nor Bob knows what k is; instead, Alice knows a number i and Bob a number j, where $k = i + j \bmod n$.

If Alice is unable to communicate, Bob must send all his answers (n bits total) to Charlie in order for Charlie to have the information he needs to pick out the right one.

Prove that just *one bit* sent from Alice to Charlie is enough to enable Bob to get away with sending only $n/2$ bits to Charlie.

Approaching Points on a Curve

The plane curve in Figure 37 has the following properties: (1) its endpoints are farther apart (using ordinary Euclidean distance in the plane) than any other pair of points on the curve; and (2) with a pencil in each hand, starting at the endpoints, you can manipulate the pencil-points along the curve until they meet, *in such a way that the (plane) distance between the two points is never increasing*.

Is there a curve with property (1) but not property (2)?

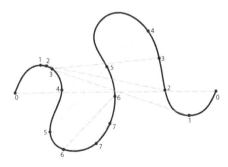

Figure 37. A "conquinculous" curve.

Sums and Products

All integers greater than one but less than 100 are put into a hat and two are drawn; Samantha is given their sum and Porfirio their product. Samantha says: "I can see that you don't know the numbers." Porfirio: "Now I do know them." Samantha: "Now I do too!"

What are the numbers?

Collapsing a Polygon

What is the minimum possible area of a simple polygon with an odd number of sides, each of unit length?

Comment: A simple polygon is a simple closed curve composed of a finite number of line segments. It need not be convex; for example, it could look like one of the polygons in Figures 38, 39, and 40. It is evident that the polygon in Figure 40 has area at least the area of a unit-side equilateral triangle, i.e. $\sqrt{3}/4$; in fact, it's not hard to see that the crown of Figure 38 is similarly constrained. Whether the "wreath" of Figure 39 can have less area than that is not so clear.

Equilateral polygons with an *even* number of sides can clearly have area as small as you want, for example by collapsing into a very sharp-pointed star. But maybe in the odd case, you can never come down to less than $\sqrt{3}/4$. Can you prove or disprove?

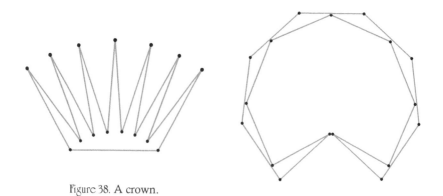

Figure 38. A crown.

Figure 39. A wreath.

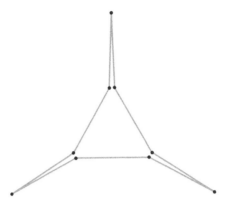

Figure 40. Collapsing to a triangle.

Sources and Solutions

Ice Cream Cake

This wonderful puzzle was passed to me by French graduate student Thierry Mora, who heard it from his prep-school teacher Thomas Lafforgue. The puzzle (of whose origin Lafforgue is unsure) actually involved a second angle as well, indicating the amount of cake passed over between wedges; it *still* requires only finitely many operations to get all the icing back on top, as ambitious readers will verify. But, the puzzle as stated here (where the second angle is 0) is already surprising and probably challenging enough.

If you thought you had a proof that infinitely many operations are necessary when x is an irrational angle, you are not alone. After all, if n operations suffice, then, the new right-side cut defining the nth wedge would have to be made at a border line between iced and un-iced areas; how could that line get there if the cake had never been cut at that point?

In fact, it can indeed get there, because when a wedge is inverted, its iced/un-iced pattern is not only complemented but *reversed*.

In analyzing this puzzle, and indeed many serious algorithmic problems as well, it helps to redefine the operation so that it is only the "state space" (here, the icing pattern on the cake), and not the operation itself, that changes from step to step. In this case, that means rotating the cake after each operation so that you are always cutting in the same place.

Accordingly, regarding "north" as $0°$, "east" as $90°$ and so forth, let us cut always at $0°$ and $-x$. The piece is then flipped over the $0°$ line to land

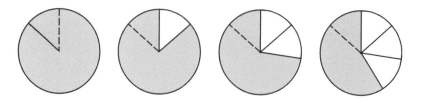

Figure 41. Cutting and flipping wedges while rotating the cake.

between $0°$ and x, while the rest of the cake is rotated clockwise by angle x. In Figure 41 the dotted lines show where the cake is to be cut for steps 1, 2, 3, and 4.

In order to see what is happening, it turns out to be easiest to think of x as a bit more than $360°/k$ for some integer k. In that case, you will cut into your first wedge during the kth operation, having encircled the cake once.

Indeed, let k be the smallest number of wedges you need to cut in order to get all the way around the cake; in other words, k is the least integer greater than or equal to $360°/x$. Then $x = y + z$, where $y = 360°/k$ and

$$0 \le z < \frac{360°}{(k-1)x} - \frac{360°}{kx} = \frac{360°}{k(k-1)x} .$$

Of course, if $z = 0$ then $x = y = 360°/k$, and it is easy to see that k operations will put all the icing on the bottom of the cake and that k more will get it back on top. Otherwise, as we shall see, you will never reach a point where all the icing is on the bottom of the cake.

As the algorithms proceed, border lines (between iced and un-iced areas) appear at angles $0, x, 2x, 3x, \ldots, (k-1)x$ and then $x - kz, 2x - kz, 3x - kz, \ldots, (k-1)x - kz$. But then they repeat. Indeed, it is easy to verify that this set S of border lines is closed under the cut-invert-replace operation. One such operation sends each ix to $(i+1)x$ except that $(k-1)x$ moves to $x - kz$; and it sends each $ix - kz$ to $(i+1)x - kz$ except that $(k-1)x - kz$ goes to x. In the meantime the line at 0 stays where it is. It follows that the set of border lines is always a subset of S.

We can already conclude that finitely many operations will return all the icing to the top, because there are only $2k - 1$ areas of cake between the $2k - 1$ lines in S. Each area can only be iced or un-iced, so the total number of cake configurations cannot exceed 2^{2k-1}. It follows that the procedure must cycle after at most 2^{2k-1} steps, but must it cycle back to the starting configuration (all areas iced)? Yes, because the operation is reversible. If it cycled back to

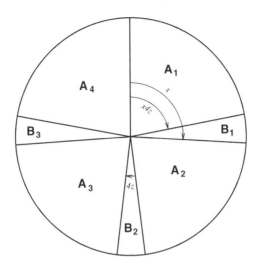

Figure 42. The set S of border lines, for $x = 93.5°$ and thus $z = 3.5°$.

some other configuration C, then there would be two different configurations which lead to C, an impossibility.

But, in fact, it is not hard to figure out exactly what happens. Between the border lines in S there are k areas of angle $x - kz$, which we will call $A_1, A_2, \ldots, A_k$, and $k - 1$ areas of angle $4z$, which we will call $B_1, B_2, \ldots, B_{k-1}$. (See Figure 42 for the case $k = 4$, $x = 93.5°$.)

As you begin cutting, the A areas turn un-iced, in order. After k operations, all are un-iced. Then they become re-iced, again in order, until after $2k$ operations they are all iced.

In the meantime, the B areas also turn un-iced in order, but since there are only $k - 1$ of those, they are all un-iced after $k - 1$ steps and all iced again after $2k - 2$ steps.

It follows that the number of steps needed to get both the A areas and the B areas fully iced is the least common multiple of the numbers $2k$ and $2k - 1$, which is $2k(k - 1)$. Thus, the complete answer is that $2k(k - 1)$ steps are required to finish the job unless $x = 360°/k$ for some integer k, in which case the B areas are non-existent and $2k$ steps suffice.

In order to get both the A and B areas *un-iced*, we would need a number n of steps that was simultaneously an odd multiple of k and an odd multiple of $k - 1$; but since exactly one of k and $k - 1$ is odd, this would require n to be both even and odd. Hence, provided the B areas exist, there is never a point when all the icing is on the bottom of the cake.

A certain very well-known mathematician's reaction, upon hearing the ice cream cake puzzle, was: "I find it hard to believe that all the icing ever returns to the top. But, one thing I'm sure of: if it does, there must be a time when it's all on the bottom!"

Hopping and Skipping

This simple-sounding puzzle was devised by mathematician James B. Shearer of IBM and appeared in the April 2007 edition of IBM's puzzle site "Ponder This."[10] In fact, it's not so easy but yields nicely to a couple of useful tricks.

Let us label the lily pads with integers in order. A natural and straightforward way to begin solving the problem is to compute the probability p that a frog starting at 1 regresses, at some point, to 0. To never regress, he must jump ahead (probability $\frac{1}{2}$) and not subsequently regress thrice (probability $1 - p^3$). Thus $1 - p = \frac{1}{2}(1 - p^3)$; divide by $(1 - p)/2$ to get $2 = 1 + p + p^2$, giving $p = (\sqrt{5} - 1)/2 \sim 0.618034$, the familiar golden ratio.

It seems awkward to try to compute the probability that the frog skips over a particular position (say, 1). You might want to start your calculations from the first time the frog hits 0, if it does, but it might already have hit 1. A better idea is to try to compute the probability that *during a particular jump* the frog leaps over a pad that he never has, and never will, hit.

For that to happen, he must (a) be jumping forward at the moment, (b) never drop back from where he lands, and (c) not have reached the pad he is jumping over in the past.

We may as well assume that the lily pad the frog is currently passing over is number 0. The key is to note that event (c) is an independent copy of event (a) if you reverse both space and time. Looking at the frog backwards in time, and regarding lower-numbered lily pads as progress, the frog appears to be behaving as before: jumping two forward or one back with equal probability. Event (c) says that when he reaches -1, he must not later "drop back" to 0.

The probability of these three events all occurring is thus $\frac{1}{2} \cdot (1-p) \cdot (1-p) = (1 - p)^2/2$. We must not forget, however, that we have not calculated the probability that 0 is skipped, only the probability that a particular jump carries the frog over a skipped pad.

Since the frog travels, on the average, at rate $\frac{1}{2}$, he is creating skipped lily pads at rate $(1-p)^2$ relative to his spatial progress. It follows that the fraction of pads he hits is $1 - (1 - p)^2 = (3\sqrt{5} - 5)/2 \sim 0.854102$.

---

[10]http://domino.research.ibm.com/Comm/wwwr_ponder.nsf/challenges/April2007.html.

Curve and Three Shadows

This puzzle was suggested to me by Rick Kenyon (University of British Columbia) who saw it on George Bergman's door at Berkeley. Bergman heard it from Hendrik Lenstra, of Berkeley and Universiteit Leiden. In Bergman's words, Lenstra had seen a toy consisting of a cubical plastic box with some slits cut in each face, forming a sort of maze in each face, with each pair of opposite faces having corresponding slits, so that a rod perpendicular to those two faces could trace out the two mazes simultaneously. But rather than one rod, there was an object that, in effect, was three mutually perpendicular rods joined at a point, one with its ends passing through the mazes on each pair of opposite faces, and was able to move around. The goal of the game was to get from some position to some other.

In fact, this puzzle is now sold commercially (for example, by the puzzle company Bits & Pieces, http://bitsandpieces.com/^Brainteasers^Metal+ Puzzles/07-W7871.html). It was invented by the brilliant Dutch puzzle designer Oskar van Deventer, whose mechanical concoctions often embody fascinating mathematical ideas.

Anyway, Lenstra observed that the maze on each face had to be a tree, since if it had a closed cycle a piece of plastic would fall out, and he wondered what the locus of positions available to the central point of the three rods was. Its projection on each face had to be a tree, but he wondered whether the set itself could have cycles. If so, then such a cycle would have to project onto each face as a tree—and thus the question was born.

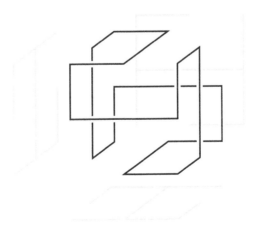

Figure 43. A closed curve with three tree-shadows.

Lenstra asked the question around February 1994, and Bergman corresponded with various people on it, with no progress. Finally, in September 1995, Kevin Buzzard, who was a postdoc at Berkeley, told Bergman and Lenstra that the question had come up earlier in Cambridge (England), and a example had been found. Buzzard had been shown the example by Imre Leader, a Cambridge University combinatorialist, who was told it by the discoverer, John Rickard. Rickard was in the mathematics department at Cambridge but now works as a computer programmer.

Rickard's example, which has a very nice six-fold symmetry, is pictured in Figure 43.

Players and Winners

This problem came from Alon Orlitsky of the University of California, San Diego. It illustrates the power of communication "from student to teacher."

Tristan and Isolde can label the teams by four-digit binary numbers, from 0000 to 1111, in alphabetical order. Then, when he finds out who played, Tristan can send Isolde 00, 01, 10, or 11, according to whether the *first position in which the two team labels differ* is the first, second, third, or fourth bit. Then Isolde simply sends back the value of that bit.

For example, if team 0011 played team 0110, and 0110 won, then Tristan would send Isolde "01" to indicate that the labels of the teams who played differed in their second bits; Isolde would send back "1" since that is the value of the second bit for the winning team.

This scheme requires that three bits be sent, only one bit less than the method in which Isolde just sends the label of the winning team. In fact, however, that one bit represents an exponential improvement! If the number of teams is $n = 2^{2^k}$, then the latter method requires 2^k bits, and the former only $k + 1$.

Charlie and the Cheaters

This puzzle, actually a serious problem in communication complexity, was considered by Harvard's Les Valiant in the '70s and communicated to me by Amit Chakrabarti of Dartmouth. Solutions for this and more general problems appear in a paper by Pavel Pudlák, Vojtěch Rödl, and Jiří Sgall [49].

Let $x_1, \ldots, x_n$ be bits representing the answers, with (say) 1 = true and 0 = false. The indices are taken modulo n. Alice sends Charlie x_{-i}; Bob sends Charlie $x_a + x_b$ for all pairs (a, b) such that $a + b = j$, where addition of bits is mod 2. Notice that there are $n/2$ such pairs (after rounding up when n is odd).

Now Charlie has x_{-i} as well as $x_{-i} + x_{i+j}$; he adds these to get x_{i+j}.

Simple, but hard to think of.

Approaching Points on a Curve

The construction of such a curve (or proof that none exists) was a challenge posed by the same Oskar van Deventer mentioned above, who was hoping to use such a curve in the design of a certain mechanical puzzle. In fact, such curves do exist; the tricky part is finding one that you can *prove* fails to have property (2).

Figure 44 shows such a curve, which, for reasons of his own, van Deventer calls *non-conquinculous*.

The purpose of the dashed circle around the curve is merely to make it easy to see that the curve satisfies property (1). To see that the curve flunks property (2), suppose otherwise and let t be the first time at which either the pencil-point beginning at the white square reaches the white arrowhead, or the point beginning at the gray square reaches the gray arrowhead. Sometime before time t, the two pencil-points must be opposed like the white and gray circles. Sometime *after* time t the points must become unopposed in order to come together, and that will cause their distance to increase.

How was this non-conquinculous curve devised? (The faint of heart should skip this and the succeeding three paragraphs.) Imagine that the curve is

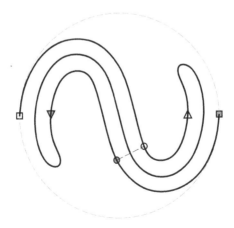

Figure 44. A non-conquinculous curve.

parametrized by a variable t; meaning that there is a continuous functions C from $[0, 1]$ to the plane such that $C(0)$ is one endpoint (say, the left), $C(1)$ is the other endpoint, and $C(t)$ traces the curve as t runs from 0 to 1.

A successful manipulation of the pencils in accordance with property (2) amounts to a pair of continuous function f and g from $[0,1]$ to itself, where at time t the pencil-points are located at $C(f(t))$ and $C(g(t))$. We thus want $f(0) = 0$, $g(0) = 1$, $f(1) = g(1)$, and at all times the plane distance between $C(f(t))$ and $C(g(t))$ to be nonincreasing as t increases. Together, f and g describe a curve from (0,1) to the line $x = y$ that stays in the triangle with vertices at (0,1), (0,0), and (1,1).

For this *not* to be possible, we would like a "dual path" from the line $x = 0$ to the line $y = 0$, avoiding the line $x = y$, consisting of pencil-point locations that are at locally minimal distance. We must then cross this path with our (f, g) curve, causing a dip in the distance.

This dual path amounts to a different kind of pencil manipulation: we start with the left pencil at the left endpoint and the right pencil somewhere else on the curve, then we move both points in the same direction along the curve until the right pencil hits the right endpoint. If we can do this in such a way that neither point can ever be moved relative to the other without increasing the distance between the two points, we have achieved our objective. In the given figure, the pencils would start at the white square and the gray arrowhead, then move together until they reach the white arrowhead and the gray square, respectively.

Your author's reward for finding a non-conquinculous curve was a prototype of the mechanical puzzle, which, like all of van Deventer's creations, is a delight.

Sums and Products

This amusing puzzle has circulated in various forms for years; it appeared in Martin Gardner's Mathematical Games column in *Scientific American* in December 1979, but was for some reason left out when that column was anthologized in *The Last Recreations* [29]. The surprise is that the vague information supplied in the puzzle statement suffices to identify the numbers.

The puzzle much as stated here was contributed independently by Steve Fenner of the University of South Carolina and by Bill Gottesman, a designer and manufacturer of sundials. The line of reasoning given below was suggested by Gottesman.

Let us begin by calling Porfirio's number P, Samantha's S, and the unknown pair $\{X, Y\}$. Since Porfirio does not at first know X and Y, X and Y

cannot both be primes, nor can they be a prime and its square or a prime and its cube. Moreover, neither can be a large prime (over 50) as it would then have to be one of the factors of P.

It follows that for Samantha to know about Porfirio's ambivalence in advance, we cannot have S larger than 53 or equal to the sum of two primes. All even numbers are out; the Goldbach conjecture, which has been verified far past 53, says that every even number greater than 2 is the sum of two primes. This leaves only numbers that exceed some odd composite by 2, namely 11, 17, 23, 27, 29, 35, 37, 41, 47, 51, and 53. Gottesman calls these numbers the "golden 11." We needn't worry about sums of a prime and its square or cube because those are all even.

From the fact that Porfirio now knows X and Y (and the fact that the sum and product of two positive integers identify them completely), we can deduce that in only one way of factoring P do the factors add up to a golden value for S. Each golden number G for Samantha yields $(G-3)/2$ possible pairs $\{X, Y\}$ (for example, 11 could be $2+9$, $3+8$, $4+7$, or $5+6$); and it must be the case that only *one* of those pairs has a product P with the desired property.

In fact, for $G = 11$ the first two pairs above would already enable Porfirio to deduce X and Y. In the $2+9$ case, $P = 18$, which factors only as 2×9 (where $2 + 9 = 11$, which is golden) or 3×6 (giving $S = 13$, which is not golden). In the $3+8$ case, $P = 24$, which factors as 2×12 (giving 14, not golden) or 3×8 (giving back 11, golden) or 4×6 (giving 10, not golden). So, G cannot be 11.

The next golden number, 17, does work! Exactly one pair of summands yields a P whose only golden sum-of-factors is 17, namely 4 and 13 (since $2 + 26 = 28$, not golden).

The rest is drudgery. We check the other six summand pairs for 17 to see that none give a viable P; this assures us that X and Y *might* be 4 and 13. To verify that $\{4, 13\}$ is the *unique* answer, we repeat the whole process for the other nine golden numbers to verify that none qualifies. This is indeed the case.

This puzzle deviates to some degree from the "elegant" solution requirement (for solved puzzles) for inclusion in this book, but the idea that a listener to this short conversation can recover the numbers without their sum *or* their product is some compensation.

Collapsing a Polygon

This puzzle was sent to me (as an unsolved problem) by Robert Veith of Southeast Indiana University, who had been seeking the answer in vain for some time. I was able to find a pretty elegant solution (I think), presented below; later I discovered that this problem and more are solved in a published paper by K. Böröczky, G. Kertész, and E. Makai, Jr., entitled "The Minimum Area of a Simple Polygon with Given Side Lengths" [9].

It turns out that the answer is: Every odd polygon with unit-length sides has area at least $\sqrt{3}/4$, with equality only for the triangle itself. How does one prove such a thing? It's certainly trivial when the polygon has only three sides, so an induction on the number of sides is a temptation. Indeed, as we will see, it's easy enough to cut a polygon with at least four sides into two polygons each with a smaller number of sides than the original. The trouble is that the new polygons will not generally be equilateral. Thus, we will need an induction hypothesis that applies to a wider class of polygons, perhaps *all* polygons.

The chosen hypothesis looks like a kludge, but works like a charm. It all depends on a certain parameter.

Let $\mathbb{O}^n$ be the set of all integer n-vectors of odd weight, i.e., $\mathbb{O}^n = \{\vec{x} = (x_1, \ldots, x_n) \in \mathbb{Z}^n \mid \sum_{i=1}^n x_i \equiv 1 \mod 2\}$. We measure how close a general polygon P is to an odd polygon with unit-length edges by means of a function $u(P)$, the *uncollapsibility* of P, defined as follows:

$$u(P) := 1 - \min_{\vec{x} \in \mathbb{O}^n} \left(\sum_{i=1}^n |e_i - x_i| \right),$$

where $e_1, \ldots, e_n$ are the lengths of the edges of P. Thus, $u(P) \leq 1$; if P is an odd polygon with unit-length edges, or indeed any polygon with integer-length edges and odd perimeter, $u(P) = 1$. On the other hand, e.g., any polygon P two of whose sides have length $\frac{1}{2}$, or whose perimeter is an even integer, has $u(P) \leq 0$.

A polygon will be deemed *proper* if it has no vertices with interior angle $180°$, that is, if its boundary bends at every vertex. If P has edges of length greater than 1, we can make from it an improper polygon P^* with all edges of length at most one by subdividing each long edge of P into edges of length 1 and at most one edge of length less than 1. Note that $u(P^*) = u(P)$.

We now proceed to prove by induction that the area $A(P)$ of any polygon is at least $\frac{\sqrt{3}}{4} u(P)$. It follows, then, that area of any polygon with an odd number of sides, each of unit length, is at least $\sqrt{3}/4$.

The key to the proof is subadditivity of the uncollapsibility function, which means the following: Let P be a polygon (not necessarily proper), and suppose that a diagonal D connecting two vertices of P divides P into polygons Q and R. Then, $u(P) \leq u(Q) + u(R)$.

To prove subadditivity, let P have edge-lengths $e_1, \ldots, e_n$, and suppose that the diagonal D has length d. Let $\vec{x} \in \mathbb{O}^n$ be such that $u(P) = 1 - \sum_{i=1}^n |e_i - x_i|$.

Let I be the set of indices of edges of P that are also edges of Q, and J the indices for the edges of R. We denote by $\vec{x} \restriction I$ and $\vec{x} \restriction J$ the restrictions of $\vec{x}$ to the edges (other than the diagonal) of Q and R, respectively; we may suppose that $\vec{x} \restriction I$ is the one with odd weight. Let d_0 be the even integer nearest d, and d_1 the odd integer nearest d, so that $|d_1 - d_0| = 1$.

Taking $\vec{x} \restriction I$ together with D-coordinate d_0 for Q, and $\vec{x} \restriction J$ together with D-coordinate d_1 for R, we have

$$
\begin{aligned}
u(R) + u(Q) &\geq 1 - \sum_{i \in I} |e_i - x_i| - |d_0 - d| + 1 - \sum_{i \in J} |e_i - x_i| - |d_1 - d| \\
&= 2 - \sum_{i \in I \cup J} |e_i - x_i| - 1 \\
&= u(P),
\end{aligned}
$$

as desired.

Next, we need to prove the main result in the case where P is a "small" triangle. Accordingly, let T be a triangle with all sides of length at most 1. Then, we will show that $A(T) \geq \frac{\sqrt{3}}{4} u(T)$, with equality only for the unit equilateral triangle.

Let the side lengths of the triangle be a, b, and c. We may assume that the integer vector yielding $u(T)$ is either $(1, 0, 0)$ or $(1, 1, 1)$; but if the former, we have $u(T) = 1 - (1 - a) - b - c < 0$ since $a < b + c$. Hence, in this case there is nothing to prove.

Otherwise, $u(T) = 1 - (1 - a) - (1 - b) - (1 - c)) = 2s - 2$, where s is the semiperimeter $\frac{1}{2}(a + b + c)$. Thus, $s > 1$ or again there is nothing to prove, and in particular $a + b$, $b + c$, and $c + a$ are all greater than 1.

We now claim that, for fixed s, the triangle T of least area is the one with two sides of length 1 (and therefore a third of size $2s - 2$).

To see this, fix a as well; by Heron's formula (a pretty proof of which can be found in [39]),

$$
\frac{A(T)^2}{s(s - a)} = (s - b)(s - c) = s - (b + c) + bc,
$$

which is minimized, since $b + c$ is constant, when $b = 1$ or $c = 1$.

Thus, we may as well permute the labels and assume that $a = 1$. But now we can repeat the argument to obtain two sides of length 1, proving the claim.

It follows that the area of T is at least the area of a $1, 1, 2s - 2$ triangle, namely

$$\sqrt{s(s-1)(s-1)(2-s)} \,.$$

But, since $s \in (1, 3/2]$, $s(2-s) \geq \frac{3}{4}$ and thus

$$A(T) \geq \sqrt{\frac{3}{4}(s-1)^2} = \frac{\sqrt{3}}{2}(s-1) = \frac{\sqrt{3}}{4}u(T)$$

with equality only for $s = 3/2$.

To prove the main theorem, we assume it is false. Then, there is a proper polygon P with $A(P) < \frac{\sqrt{3}}{4}u(P)$ that has a minimum number n of edges. If $n = 3$, we order triangles lexicographically according to $(\lceil c \rceil, \lceil b \rceil, \lceil a \rceil)$, where $a \leq b \leq c$ are the side lengths of P, and require that P be minimal in this (partial) order.

Assume first that $n > 3$, and let D be any diagonal from one vertex of P to another. Such a diagonal must exist because if P is convex, any two nonadjacent vertices are connected by an interior line segment. Otherwise, there is a vertex v with interior angle greater than π, and if we scan the interior of P from v, starting in the direction of one of the incident edges and ending at the other, we must see more than one other edge; where the scan passes from one such edge to another, there is a vertex to which we can connect v by a diagonal.

The diagonal splits P into proper polygons Q and R, each with fewer than n edges and hence each satisfying the inequality of the theorem. But then by subadditivity, we have

$$A(P) < \frac{\sqrt{3}}{4}u(P) \leq \frac{\sqrt{3}}{4}(u(Q) + u(R)) \leq A(Q) + A(R) = A(P),$$

a contradiction.

Hence we are reduced to the case $n = 3$; let A be the vertex opposite a, etc. We know from the small-triangle analysis that $\lceil c \rceil > 1$. If $\lceil b \rceil < \lceil c \rceil$ we draw a diagonal from C to any new vertex of P^*; the length of this diagonal is less than b, since the angles incident to the long side are acute. Then, the two triangles into which the diagonal splits P^* lie below P in the lexicographic order, thus subadditivity together with the triangle analysis gives us a contradiction.

If $\lceil b \rceil = \lceil c \rceil > 1$, we chose P^* so that there is a new vertex U (respectively V) of P^* on side b (respectively c) at distance 1 from vertex C (respectively B). Now we draw *two* diagonals, one from U to V (of length d) and another from V to C (of length e). Again applying subadditivity we deduce that one of the three small triangles BCV, CVU, and VUA thus created must be a counterexample to the theorem. However, all three precede P in the inductive ordering, and this final contradiction proves the theorem.

Observe that the induction, combined with the strictness condition in the small-triangle assertion, shows that the inequality of the main theorem is strict for any non-degenerate polygon P, unless P is precisely an equilateral triangle with edges of unit length.

Whew!

Unsolved and Just-Solved

Wir müssen wissen. Wir werden wissen.

—David Hilbert (1862–1943)

It's not an accident that this is Chapter 11: this one could break you. The good news is that unsolved is by no means the same as unsolvable, and in fact two of the "unsolved puzzles" from my previous book have recently been solved. The first was a particularly notorious puzzle, and what happened with it is truly bizarre.

Conway's Angel and Devil

An Angel flies over an infinite checkerboard, and every now and then she must alight on a square. She can travel no more than 1000 King-moves in the air before she lands.

While she's in the air, however, the Devil, who lives below the board, can destroy any one square of his choice.

Can the Devil trap the Angel?

In stunning and inexplicable fashion, this puzzle—open for thirty years—was suddenly solved, independently and almost simultaneously, by *four* people in four different countries [10, 20, 40, 43].

The methods were in most cases similar, but they did not rely on some recently-discovered new technique; in fact, they began instead with observations made by John H. Conway himself in the '70s. The four solvers were András Máthé, of Eötvös Loránd University in Budapest; Brian Bowditch, of the University of Southampton; Oddvar Kloster, of SINTEF ICT in Oslo; and Peter Gács, of Boston University. It was long known that the angel of power 1 (that is, who can move only one King-step when she flies) could be defeated. Máthé and Kloster showed that the angel of power 2 can win the game; Bowditch proved power 4 is enough, and Gács showed that *some* power is enough.

The proofs were simple enough that Béla Bóllobás, of Cambridge University and the University of Memphis, was able to describe an amalgamation of them in a delightful one-hour talk at the University of Illinois. What follows is an informal description, based on his talk and Máthé's paper, of a proof that the angel of power 5 can win.

The general idea is to show that if the angel can win (in a slightly stronger sense) against a constrained adversary called the "nice devil," then she can win against the original devil. And it turns out that she has a surprisingly simple winning strategy against the nice devil.

The nice devil is not allowed to eat a square to which the angel could previously have jumped; putting it another way, all squares at distance between 0 and p from a square previously visited by the angel are off limits to the nice devil. It is easy to verify that the angel can win the original game if, for any n, she can escape to distance n; we now show that if she can do this for the nice devil, she can do it against the original devil.

To see this, we assume that the original devil *does* have a strategy to keep the angel within distance n of the origin, and we show how the nice devil can do the same. Given any sequence of moves by the angel, we create a "reduced" sequence as follows. Let A_1 be the earliest-visited place from which she could have jumped directly to her ending point A_0; delete all moves between A_1 and A_0. Now, let A_2 be the earliest-visited place from which she could have jumped directly to A_1, and delete all moves between A_2 and A_1. Continuing in this way we get an abridged version $A_k, A_{k-1}, \ldots, A_1, A_0$ of the original sequence, in which the angel never jumps to a point to which she could have moved earlier.

We now direct the nice devil to respond to a given sequence of moves the way the original devil would have responded to the reduced sequence, except that if this asks him to eat an impermissible square (or one he has already eaten), he eats any permissable one instead. It is easy to show that if a given sequence is possible for the angel against the nice devil—that is, it doesn't land her on an eaten square—then the reduced version of that sequence works against the original devil. Thus, if she can escape to distance n against the nice devil, she can do the same against the original devil and thus win the game.

We have reduced the problem to escaping the nice devil, but that turns out to be remarkably easy even if we permit the angel only to run, not jump, along uneaten squares. She starts at the square whose lower left-hand corner is at the origin and imagines a wall along the line $y = 0$. Every time the nice devil eats a square, she erects a wall around that square. In the meantime, at each turn, she runs generally northward keeping the wall immediately at her left, as far as she is allowed. Occasionally she will have to run southward to get around some section of wall, but power 5 is enough to ensure that she will have progressed northward at least one unit for each step she has taken. It follows that the angel of power 5 can escape to any arbitrary distance, and with more work, one can show that already power 2 is enough. Figure 45 shows a possible path for the angel of power 2.

It is worth noting that this strategy for the angel will most assuredly *not* work against the original devil, who, for example, can set a trap for the angel far up the *y*-axis in which she is lured to the end of a peninsula surrounded by

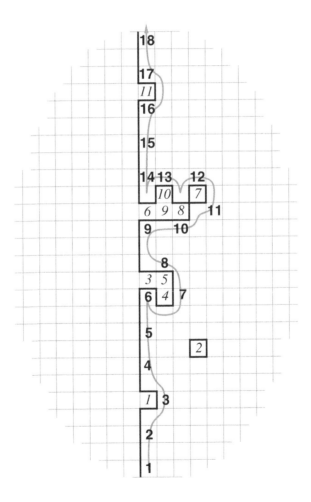

Figure 45. The wall (in black), angel (boldface), and nice devil (italics).

a vast sea of eaten squares, then cut off from the shore. The nice devil can't do this because he is not permitted to eat away the base of the peninsula after the angel has passed that way.

It's unfortunately rather hard to explain how the running strategy morphs into a successful strategy against the original devil, even though the reduction above seems straightforward. This explains to some extent why the puzzle remained open so long and reinforces the observation that reductions can be a powerful way to solve a problem.

The second puzzle has a long way to go to be completely solved, but until recently *nothing* had been proved.

Gridlock

Vertices of the infinite plane grid are chosen independently with probability $p \in (0, 1)$, and according to fair coin flips, each chosen vertex is occupied either by a car facing north or a car facing east.

The cars are controlled by a traffic signal that alternates "green-east" and "green-north." When it turns green-east, each eastbound car whose right-hand neighbor vertex is unoccupied moves to that vertex; the others (including those blocked by another eastbound car) remain where they are. When the signal turns green-north, each unblocked northbound car moves one vertex northward.

Experiments suggest that when p is below a certain critical value p_0, the cars gradually break free; that is, each has a limiting velocity equal to the velocity of a car that is never blocked. But when $p > p_0$, the opposite occurs: The cars get hopelessly tangled and every car takes only finitely many moves before being blocked forever. Now, can any of this be proved?

This model of traffic flow at the intersection of two major one-way streets was introduced by O. Biham, A. A. Middleton, and D. Levine in 1992 [6]. Its bizarre behavior has attracted much interest; you can find a bibliography at http://cui.unige.ch/spc/Bibliography/traffic.html.

Pictured in Figure 46 are finite pieces of a free configuration and of a gridlocked configuration, each somewhat typical of what has appeared in experiments run by Raissa D'Souza [15], now of the University of California at Davis.

In spring 2005, at the Mathematical Sciences Research Institute in Berkeley, Omer Angel, Ander Holroyd, and James Martin finally made some progress: they proved that the jammed phase really does exist. In other words, they showed that at sufficiently high density the cars do jam so that every car

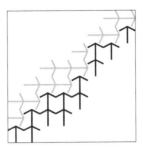

Figure 46. A free configuration (left) and a gridlocked configuration (right).

ends up taking only finitely many steps. Their proof involves the theory of percolation and will not be repeated here, but it is very clever and interested readers are encouraged to read it. The paper is called "The Jammed Phase of the Biham-Middleton-Levine Traffic Model" [2].

Of course, in mathematics every time a problem is solved, three new ones rear up in its place. Here are a few unsolved beauties that I think deserve some attention.

Packing Rectangles

Suppose that you are given a finite set of points in a square, including the lower left corner. Can you construct a set of disjoint rectangles in the square, each having as its lower left-hand corner one of the given points, whose total area is at least half the area of the square?

This curious and frustrating problem was communicated to me more than ten years ago by Bill Pulleyblank (mathematician and executive) of IBM, who did not remember where he had heard it. Since then the problem has reappeared from time to time, but I have not been able to trace it back to any earlier source than Bill himself. It did appear in June 2004 on IBM's own puzzle webpage called "Ponder This,"[11] but remains unsolved; I do not even know of a proof that the rectangles can be persuaded to cover 1% of the square's area.

Pictured in Figure 47 is a configuration of points together with a satisfactory set of rectangles.

---

[11] http://domino.research.ibm.com/Comm/wwwr_ponder.nsf/challenges/June2004.html.

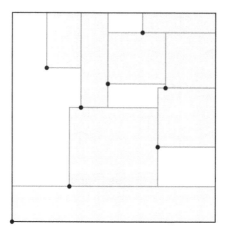

Figure 47. A successful covering of more than half the square, from eight given points.

Products and Sums

Can you color the natural numbers $\{0, 1, 2, \dots\}$ with finitely many colors, in such a way that the sum $x + y$ and the product xy of any two whole numbers always have different colors?

This problem was brought to my attention by David Galvin, a postdoc at the University of Pennsylvania. Apparently a set of six numbers is known such that any two of them are the product and sum of some pair of numbers, so at least six colors would be needed; on the other hand, there are not arbitrarily large sets of numbers with the above property.

The next puzzle involves a game called "Peer Pressure" invented by Boris Alexeev of the University of Georgia, which proved popular with a recent USA Mathematics Olympiad Team.

Peer Pressure

Two players are dealt some number of cards, initially *face up*, each card carrying a different integer. In each round, the players simultaneously play a card; the higher card is discarded and the lower card passed to the other player. The player who runs out of cards loses.

As the number of cards dealt becomes larger, what is the limiting probability that one of the players will have a winning strategy?

Boris suspects (as do I) that this probability goes to zero, but it seems to be difficult to analyze this simple variant of the game "War."

Here's a disturbing quickie from Steve Hedetniemi of Clemson University:

Coverage by Queens

Let $f(n)$ be the minimum number of queens needed on an $n \times n$ chessboard so that every square is occupied or covered by a queen. Is it always the case that $f(n + 1) \geq f(n)$?

There are many puzzles concerning the placement of chess pieces (usually queens or knights) on an $n \times n$ board; with queens the object is usually to place as many as possible so that no queen attacks another, but here the idea is to use few queens and cover every square. It seems hard to believe that you might need more queens to cover fewer squares, but since a larger board gives you more places from which a queen can oversee her dominion, it's hard to rule that out!

Here's an entertaining but actually pretty serious puzzle that has confounded optimization experts for years.

Rendezvous

Two friends become separated while shopping at the Mall of America. It takes 15 minutes to search a shop but negligible time to get from one shop to another (the mall is well organized in a giant multi-story square). Not having arranged a meeting point, or decided in advance who should search and who should stay put, what should they do to minimize the expected time to find each other?

If one friend searched while the other stayed fixed, it would take on the average $n/2$ store-searches to meet, where n is the (large) number number of stores in the Mall of America. However, the "rules" of *rendezvous* forbid a protocol that breaks the symmetry of the two friends, i.e., by saying the younger one searches while the older sits. If *both* friends search, it takes n steps on average before they get lucky and find themselves searching the same store at the same time.

The rendezvous problem was posed in 1976 (not in these words) by Steve Alpern of the London School of Economics, and you can read about it and some other problems on Richard Weber's website[12] at Cambridge University.

---

[12] http://www.statslab.cam.ac.uk/~rrw1/talks/weber-k3-seminar.pdf.

Weber and E. J. Anderson have proposed an algorithm whereby each friend flips a bent coin, deciding with probability about 0.2475 to stay put and otherwise search the n stores in random order; if nothing happens, repeat the whole process. This succeeds in an average of $0.8289n$ steps, but no one has been able to prove it is optimal.

Twisting the Rectangle

As you probably know, you can make a Möbius band out of a long rectangular strip of paper by giving the strip a half-twist and joining the ends. How long a strip do you need? In other words, what are the proportions of the squarest rectangle of paper from which you can make a Möbius band, without stretching or creasing?

In Lecture 14 of their book, *Mathematical Omnibus: Thirty Lectures on Classic Mathematics* [19], Dmitry Fuchs and Serge Tabachnikov present this puzzle along with a proof that a length/width ratio of $\pi/2 \sim 1.57$ is necessary and $\sqrt{3} \sim 1.83$ is sufficient. But, the precise answer is unknown.

Gumball Machines

The various gumball machines at the local arcade work randomly, sometimes producing many gumballs and sometimes none at all. But, each machine gives an average of one gumball when it is operated. What is the maximum possible probability that if all n machines are operated simultaneously, more than n gumballs will result?

This puzzle (stated in terms of independent random variables) is due to Uri Feige of Microsoft Research. It looks as if the best you can do is to have each machine produce $n + 1$ gumballs with probability $1/(n + 1)$, and none otherwise; then, as long as at least one machine cooperates, you will get more than n gumballs. This happens with probability

$$1 - \left(1 - \frac{1}{n+1}\right)^n,$$

which gets as close as you want to $1 - e \sim 63\%$ as n gets large. No one has found any better way to do it, but the best bound anyone has achieved is Feige's own, that you can never get the probability of more than n gumballs to exceed 12/13. Seems like this shouldn't be so hard.

Discs on the Plane

Suppose that a collection of open unit discs is given which is a thousand-fold cover of the plane; that is, every point of $\mathbb{R}^2$ is covered by at least 1000 discs. Prove that you can color each disc red or blue in such a way that the red discs and blue discs each cover the plane.

János Pach, of New York University, is the originator of (and expert on) this wonderful open problem. In his paper "Covering the plane with convex polygons" [46], Pach proves that, for any symmetric polygon P and any positive integer r, there is a k such that any k-fold covering of the plane by translates of P can be partitioned into r covers. But, no such k is known when the polygon becomes a disc, even for $r = 2$.

Personally, I think $k = 4$ ought to be enough. What do you think?

Afterword

I hope that posterity will judge me kindly, not only as
to the things which I have explained, but also to those
which I have intentionally omitted so as to leave to others
the pleasure of discovery.

—René Descartes (1596–1650), *La Geometrie*

You may or may not believe Descartes, but you certainly should not believe
me if I told you I deliberately left out stuff for you to discover on your own.
But, there are certainly *lots* of great puzzles missing from this volume that
are worthy of your attention, and many more yet to be devised. The sources
mentioned in this book, many of them online, are a great place to look; and
you may come up with some beauties on your own.

Puzzles are not a substitute for learning mathematics by traditional methods, but they will help you remember the math you *did* learn, and they can
surely entertain and exercise the mind. The ones in this volume are selected
with an additional purpose, as well: to keep you from becoming complacent
about your mathematical intuition.

It works for me, anyway.

Peter Winkler
February 28, 2007

Bibliography

[1] G. Aggarwal, A. Fiat, A. V. Goldberg, J. Hartline, N. Immorlica, and M. Sudan (2005). "Derandomization of Auctions," in *Proceedings of the 37th Annual ACM Symposium on Theory of Computing*, ACM Press, 619–625.

[2] O. Angel, A. E. Holroyd, and J. B. Martin (2005). "The Jammed Phase of the Biham-Middleton-Levine Traffic Model." *Elec. Commun. in Probability* 10, Paper 17, 167–178.

[3] E. Berlekamp and J. P. Buhler (1999–2007). "Puzzle Column." *Emissary* (Newsletter of the Mathematical Sciences Research Institute, Berkeley, CA).

[4] E. R. Berlekamp, J. H. Conway, and R. K. Guy (2001–2004). *Winning Ways for Your Mathematical Plays*, Volumes I, II, III, and IV. A K Peters, Ltd.

[5] E. R. Berlekamp and T. Rodgers, editors (1999). *The Mathemagician and Pied Puzzler*, A K Peters, Ltd.

[6] O. Biham, A. A. Middleton, and D. Levine (1992). "Self-Organization and a Dynamical Transition in Traffic-Flow Models." *Phys. Rev. A* 46:10, R6124–R6127.

[7] H. Boerner (1955). *Darstellungen von Gruppen*, Springer-Verlag. (Second edition 1967.)

[8] E. D. Bolker and H. Crapo (1979). "Bracing Rectangular Frameworks I." *SIAM J. Appl. Math.* 36:3, 473–490.

[9] K. Böröczky, G. Kertész, and E. Makai, Jr. (1999). "The Minimum Area of a Simple Polygon with Given Side Lengths." *Periodica Mathematica Hungarica* 39:1–3, 33–49.

[10] B. H. Bowditch (2007). "The Angel Game in the Plane." *Comb., Prob. and Comput.* 16:3, 345–362.

[11] B. Brown (1933). "Solution to Problem E36." *Amer. Math. Monthly* 40, 607.

[12] J. G. Coffin (1923). "Problem 3009." *Amer. Math. Monthly* 30:2, 76.

[13] E. Curtin and M. Warshauer (2006). "The Locker Puzzle." *The Mathematical Intelligencer* 28:1, 28–31.

[14] P. Diaconis, R. L. Graham, and B. Sturmfels (1996). "Primitive Partition Identities," in *Combinatorics: Paul Erdos is Eighty*, Volume II, Bolyai Society Mathematical Studies 2, János Bolyai Mathematical Society, 173–192.

[15] R. M. D'Souza (2005). "Coexisting Phases and Lattice Dependence of a Cellular Automata Model for Traffic Flow." *Phys. Rev. E* 71, 0066112.

[16] S. J. Einhown and I. J. Schoenberg (1985). Problem 3a in "Puzzle Section." *Pi Mu Epsilon Journal* 8:3, p. 178.

[17] D. Fomin (1990). *Zadaqi Leningradskih Matematitcheskih Olimpiad*, Leningrad.

[18] A. Friedland (1971). *Puzzles in Math and Logic*, Dover.

[19] D. Fuchs and S. Tabachnikov (2007). *Mathematical Omnibus: Thirty Lectures on Classic Mathematics*, draft. Available at http://www.math.psu.edu/tabachni/Books/taba.pdf.

[20] P. Gács (2007). "The Angel Wins," preprint. Available at http://www.cs.bu.edu/~gacs/papers/angel.pdf.

[21] A. Gal and P. B. Miltersen (2003). "The Cell Probe Complexity of Succinct Data Structures," in *Automata, Languages and Programming: 30th International Colloquium, ICALP 2003, Eindhoven, The Netherlands, June 30–July 4, 2003, Proceedings*, Lecture Notes in Computer Science 2719, Springer, 332–344.

[22] J. A. Gallian and D. J. Rusin (1979). "Cyclotomic Polynomials and Nonstandard Dice." *Disc. Math.* 27, 245–259.

[23] G. A. Galperin and A. K. Tolpygo (1986). *Moscow Mathematical Olympiads*, Prosveshchenie.

[24] M. Gardner (1977). *Mathematical Carnival*, Vintage.

[25] M. Gardner (1978). "Mathematical Games." *Scientific American* 238, 19–32.

[26] M. Gardner (1971). *The Sixth Book of Mathematical Puzzles and Diversions from Scientific American*, Simon & Schuster.

[27] M. Gardner (1988). *Hexaflexagons and Other Mathematical Diversions: The First Scientific American Book of Puzzles and Games* (reprint edition), University of Chicago Press.

[28] M. Gardner (1989). *Penrose Tiles to Trapdoor Ciphers*, W. H. Freeman & Co.

[29] M. Gardner (1997). *Last Recreations: Hydras, Eggs, and Other Mathematical Mystifications*, Springer Verlag.

[30] M. Gardner (2005). *The Colossal Book of Short Puzzles and Problems*, W. W. Norton & Co.

[31] E. Goles and J. Olivos (1980). "Periodic Behavior of Generalized Threshold Functions." *Disc. Math.* 30, 187–189.

[32] O. Gossner, P. Hernández, and A. Neyman (2007). "Online Matching Pennies." Available at http://ratio.huji.ac.il/dp/dp316.pdf.

[33] N. Goyal, S. Lodha, and S. Muthukrishnan (2006). "The Graham-Knowlton Problem Revisited." *Theory Comput. Syst.* 39:3, 399–412.

[34] E. Gutkin (2005). "Blocking of Billiard Orbits and Security for Polygons and Flat Surfaces." *Geom. and Funct. Anal.* 15, 83–105.

[35] J. F. Hall (2005). "Fun with Stacking Blocks." *Amer. J. Phys.* 73:12, 1107–1116.

[36] C. Hardin and A. D. Taylor (2007). "A Peculiar Connection between the Axiom of Choice and Predicting the Future." *Amer. Math. Monthly*, to appear.

[37] G. H. Hardy (1907). "On Certain Oscillating Series." *Quart. J. Math.* 38, 269–288.

[38] C. P. Jargodzki and F. Potter (2001). "Challenge 271: A Staircase to Infinity," in *Mad About Physics: Braintwisters, Paradoxes, and Curiosities*, Wiley, 246.

[39] D. A. Klain (2004). "An Intuitive Derivation of Heron's Formula." *Amer. Math. Monthly* 111:8, 709–712.

[40] O. Kloster (2007). "A Solution to the Angel Problem," preprint. Available at http://home.broadpark.no/~oddvark/angel/Angel.pdf.

[41] D. E. Knuth (1998). *The Art of Computer Programming*, Volume 3: Sorting and Searching, second edition, Addison-Wesley.

[42] J. D. E. Konhauser, D. Velleman, and S. Wagon (1996). *Which Way Did the Bicycle Go*, Mathematical Association of America.

[43] A. Máthé (2007). "The Angel of Power 2 Wins." *Comb., Prob. and Comput.* 16:3, 363–374.

[44] Ş. Nacu and Y. Peres (2005). "Fast Simulation of New Coins from Old." *Ann. Appl. Probab.* 15:1A, 93–115.

[45] B. E. Oakley and R. L. Perry (1965). "A Sampling Process." *The Mathematical Gazette* 49:367, 42–44.

[46] J. Pach (1986). "Covering the Plane with Convex Polygons." *Discr. Comput. Geom.* 1:1, 73–81.

[47] M. Paterson and U. Zwick (2006). "Overhang," in *Proceedings of the 17th Annual ACM-SIAM Symposium on Discrete Algorithms*, ACM Press, 231–240. (Full article to appear in *Amer. Math. Monthly*.)

[48] M. Paterson, Y. Peres, M. Thorup, P. Winkler, and U. Zwick (2007). "Maximum Overhang," preprint.

[49] P. Pudlák, V. Rödl, and J. Sgall (1997). "Boolean Circuits, Tensor Ranks and Communication Complexity." *SICOMP* 26:3, 605–633.

[50] S. Robinson (2001). "Why Mathematicians Now Care about Their Hat Color." *New York Times* April 10, http://www.msri.org/people/members/sara/articles/hat.html.

[51] D. O. Shklarsky, N. N. Chentov, and I. M. Yaglom (1962). *The USSR Problem Book*, W. H. Freeman and Co.

[52] I. J. Schoenberg (1982). *Mathematical Time Exposures*, Mathematical Association of America.

[53] S. Singh (2000). *The Code Book: The Science of Secrecy from Ancient Egypt to Quantum Cryptography*, Anchor Press.

[54] R. Sprague (1963). *Recreation in Mathematics*, Blackie & Son Ltd.

[55] R. M. Smullyan (1978). *What is the Name of This Book: The Riddle of Dracula and Other Logical Puzzles*, Prentiss-Hall.

[56] S. Tabachnikov (2005). *Geometry and Billiards*, American Mathematical Society.

[57] B. Tenner (2004). "A Non-Messing-Up Phenomenon for Posets." Available at http://arxiv.org/abs/math.CO/0404396.

[58] C. Wang (1993). *Sense and Nonsense of Statistical Inference*, Marcel Dekker.

[59] P. Winkler (2004). *Mathematical Puzzles: A Connoisseur's Collection*, A K Peters, Ltd.

[60] W. A. Wythoff (1907). "A Modification of the Game of Nim." *Nieuw Arch. Wiskunde* 8, 199–202.

[61] N. Yoshigahara (2003). *Puzzles 101: A Puzzlemaster's Challenge*, A K Peters, Ltd.

Index of Puzzles

Praise for

Mathematical Puzzles: A Connoisseur's Collection

by Peter Winkler

"Beyond the excellent selection, the reader will appreciate the author's exquisite labor to state each puzzle perfectly. The solutions, when one has need of them, read just as well."

—*Choice* magazine

"Winkler's book will certainly appeal to the mathematician, as well as to students of all ages—high-school, college, and graduate. His philosophy of what constitutes a good puzzle is right on the mark, showing that this Connoisseur's Collection really is of quality and depth."

—*Read This!*, MAA Online

Collected over several years by Peter Winkler of Bell Labs, dozens of elegant, intriguing challenges are presented in *Mathematical Puzzles*. The answers are easy to explain, but without this book, devilishly hard to find. Creative reasoning is the key to these puzzles. No involved computation or higher mathematics is necessary, but your ability to construct a mathematical proof will be severely tested—even if you are a professional mathematician. For the truly adventurous, there is even a chapter on unsolved puzzles.

Other recreational mathematics titles available from A K Peters

Origami Design Secrets: Mathematical Methods for an Ancient Art, Robert J. Lang

Marvelous Modular Origami, Meenakshi Mukerji

Geometric Puzzle Design, Stewart Coffin

Piano-Hinged Dissections: Time to Fold!, Greg Frederickson

Puzzles 101: A Puzzlemasters Challenge, Nobuyuki Yoshigahara

Luck, Logic, and White Lies: The Mathematics of Games, Jörg Bewersdorff

More related titles online at www.akpeters.com